North American Mixing Forum

Core Programming Topic at the 2011 AIChE Annual Meeting

Minneapolis, Minnesota, USA
16-21 October 2011

ISBN: 978-1-61839-727-0

Printed from e-media with permission by:

Curran Associates, Inc.
57 Morehouse Lane
Red Hook, NY 12571

Some format issues inherent in the e-media version may also appear in this print version.

Printed by Curran Associates, Inc. (2012)

For permission requests, please contact AIChE
at the address below.

AIChE
3 Park Avenue
New York, NY 10016-5991

Phone: (203) 702-7660
Fax: (203) 775-5177

www.aiche.org

Additional copies of this publication are available from:

Curran Associates, Inc.
57 Morehouse Lane
Red Hook, NY 12571 USA
Phone: 845-758-0400
Fax: 845-758-2634
Email: curran@proceedings.com
Web: www.proceedings.com

TABLE OF CONTENTS

(36a) Pilot-Scale Testing of Solid Suspension by Rotating Jets In a Tank

Michael Poirier and Zafar Qureshi, Savannah River National Laboratory, Aiken, SC

The Small Column Ion Exchange (SCIX) process is being developed to remove cesium, strontium, and select actinides from Savannah River Site (SRS) Liquid Waste using an existing waste tank (85 ft diameter) to house the process. Savannah River National Laboratory (SRNL) is conducting pilot-scale mixing tests to determine the pump requirements for suspending monosodium titanate (MST), crystalline silicotitanate (CST), and simulated sludge. The purpose of this pilot scale testing is to determine the requirements for the pumps to suspend the MST particles so that they can contact the strontium and actinides in the liquid and be removed from the tank.

The pilot-scale tank is a 1/10.85 linear scaled model of an SRS waste tank. The tank diameter, tank liquid level, pump nozzle diameter, pump elevation, and cooling coil diameter are all 1/10.85 of their full-scale dimensions. The pump locations correspond to the proposed locations in the waste tank. This presentation will discuss the testing and results.

See more of this Session: Mixing In Multi-Phase Systems I
See more of this Group/Topical: North American Mixing Forum

(36b) Suspension of Solid Mixtures by Mechanical Agitation

Kevin J. Myers[1], Tianxin Bao[1], Eric E. Janz[2] and Julian B. Fasano[3], (1)Chemical and Materials Engineering, University of Dayton, Dayton, OH, (2)Chemineer, Inc., Dayton, OH, (3)Mixer Engineering Company, Troy, OH

The suspension of solids using mechanical agitators has been widely studied, resulting in methods that permit reasonably accurate agitator design for suspending particles with uniform characteristics (size, shape, density, etc.). However, limited work has been performed in the area of suspending mixtures of solids with different characteristics, but there has been a growing interest in this subject. An experimental program has been undertaken to develop guidelines that can be used to design agitators for the suspension of solid mixtures. Although this work is ongoing, the following results have been obtained.

- The just-suspended speed of some solid mixtures is equal to the speed required to suspend the more difficult to suspend solid alone (i.e. – addition of the easier to suspend solid does not alter the just-suspended speed).

- The power required to suspend a solid mixture can be less than, equal to, or greater than the sum of the powers required for suspension of the individual solids.

Given these findings, there appears to be no simple rule to predict the just-suspended speed of a mixture of solids from the just-suspended speeds of the individual solids. This presentation will review the experimental work that has been performed with an aim to developing agitator design rules for suspending solid mixtures.

See more of this Session: Mixing In Multi-Phase Systems I
See more of this Group/Topical: North American Mixing Forum

CHARACTERIZATION OF LIQUID-PULP FIBER SUSPENSION MIXING AT TEE MIXERS BASED ON ELECTRICAL RESISTANCE TOMOGRAPHY

Wisarn Yenjaichon, John R. Grace, Jim C. Lim and Chad P.J. Bennington⁺*
Department of Chemical and Biological Engineering
University of British Columbia
Vancouver, BC, Canada V6T1Z3

Introduction

Mixing is an essential unit operation in the pulp and paper industry. In bleach plants, pre-distribution of chemicals in pulp suspensions before entering tower reactors is very important, and in-line mixers have normally been used to ensure efficient contacting between chemicals and pulp to achieve the greatest lignin removal and optimal use of the bleaching agent. Chemical injection in pulp mixing operations is commonly used as a pre-distributor to mix liquid chemicals into pulp suspensions ahead of various mixers including, static mixers, peg mixers and high-shear mixers. The simplest type is the tee mixer. Understanding jet mixing behavior at a tee-junction provides basic concepts for further analysis of different types of in-line mixers. However, few studies have been reported on the in-line mixing behavior of liquid injection into pulp suspensions (Luettgen et al., 1991; Rewatkar et al., 2002). For our current project, a non-invasive technique "electrical resistance tomography (ERT) has been used to study mixing pulp fiber suspensions at tee-mixers for various operating conditions to provide better understanding of jet mixing in pulp suspension flow.

Experimental

The experiments were conducted in a pilot-scale flow loop facility. Mixing quality of liquid injection into the main stream was evaluated after 90° tee-mixers for both water and softwood kraft pulp suspensions over a range of mass concentrations (C_m = 0 – 3%), main stream or pipe velocities (U_p = 0.5 – 5 m/s), side stream or jet velocities (U_j = 1 – 12.5 m/s) and jet-to-pipe diameter ratio (D_r = 0.05 – 0.208). The acrylic test section was installed 4.5 meters downstream of the pipe bend, ensuring that the flow is fully developed before entering the test section. The injection tube length was sufficiently long to provide fully developed turbulent flow before combining with the main stream. For all tests in this work, the fluid temperature in the main stream was kept at 20±5°C and the temperature difference between the main and side stream was kept below 10 °C. The degree of mixing was determined from measurements in cross-sectional planes along the pipeline based on a modified mixing index derived from the coefficient of variation (*CoV*) of individual conductivity values in each image pixel.

Results and Discussion

Mixing Quality in Newtonian Fluid (Water)

All tests were in the turbulent flow regime for both the main stream (Re_p = 38,100 – 381,000) and the jet stream (Re_j = 9,750 – 48,600). Figure 1 shows the effect of main

* *Author to whom correspondence may be addressed.*
E-mail address: wyenjaichon@chbe.ubc.ca
⁺ *Deceased*

stream velocity on the mixing quality for the same jet-to-pipe velocity ratio and diameter ratio. The mixing quality was almost independent of the main stream velocity. The small difference was likely due to the retention time. The mixing quality was slightly better when the flow rate was lower.

For tee-mixers, the mixing quality of Newtonian fluids in turbulent flow depends strongly on jet penetration and thus on the jet-to-pipe velocity ratio. For jet penetration in fully developed turbulent pipe flow, three mixing modes were defined by Pan and Meng (2001): wall-source, jet-mixing and jet-impaction. Figure 2 illustrates the effect of jet-to-pipe velocity ratio on the mixing quality at a jet-to-pipe diameter ratio, D_r, of 0.05. The mixing quality significantly improved as the jet-to-pipe velocity ratio increased. At low velocity ratios ($R \leq 3$), the jet momentum was insufficient for the jet stream to escape from the pipe wall, and it therefore became sidewall injection (i.e. wall-source mode). The mixing quality was then very poor. At higher velocity ratios ($4 \leq R \leq 10$), the jet momentum was sufficient for the jet to escape from the wall region and join the main flow. The modified mixing index was significantly lower, i.e. the mixing quality was much better for the jet-mixing mode. Further increases in velocity ratio ($R > 10$) allowed the jet to impinge on the far wall of the pipe. The jet then split and travelled around the pipe periphery, with mixing being efficient downstream due to the breakdown of large vortical structures into smaller eddies from the jet impaction (Tosun, 1987). The jet penetrations in each scenario are shown in the tomographic images in Figure 3.

Mixing Quality in Pulp Fiber Suspensions

The results show significant differences in jet mixing behavior of non-Newtonian pulp suspensions and the Newtonian fluid (water). For pulp fiber suspensions, the mixing quality depends strongly on the flow regime, which varies with the main stream velocity. It is customary to identify the flow regime as plug, mixed or turbulent flow based on criteria proposed by Robertson and Mason (1957). At very low velocity, the suspension travels as a plug (referred to as "plug flow") with an apparent fiber-wall friction. As the velocity increases, an annular water layer in laminar flow occurs around the plug. Further increases in velocity produce turbulent stresses in the water layer causing the plug to disintegrate. At higher velocity, the plug continues to be disrupted until it disappears and the flow becomes fully turbulent.

Figure 4 shows the effect of the main stream velocity on the mixing quality for pulp suspension at C_m = 0.5% at almost the same jet-to-pipe velocity ratio and the same diameter ratio. The effect of fiber network on the mixing quality can be significant, even at a mass concentration as low as 0.5%. At low velocities ($U_p \leq 1$ m/s), plug flow was approached and mixing was very poor due to the strong fiber network preventing tracer distribution downstream. At higher velocities ($1 < U_p \leq 3$ m/s), the plugs disintegrated, and the mixing quality significantly improved downstream. Further increases in velocity ($U_p > 3$ m/s) continued to disrupt the plug until the flow reached the turbulent regime, where the mixing quality approached that for the Newtonian fluid (water).

At a higher mass concentrations ($C_m \geq 2\%$), mixing was poor, even at a high mainstream velocity, since the turbulent shear was not high enough to disrupt the plug, and the flow remained in the plug flow regime. Figure 5 illustrates the effect of the mainstream velocity on the degree of mixing for a pulp suspension at C_m = 2.0% at almost the same jet-to-pipe velocity ratio and the same diameter ratio. The mixing quality slightly improved as the mainstream velocity increased. However, significantly lower mixing quality was observed compared to the mixing quality of water (C_m = 0%) pipe flow.

Conclusions

In-line jet mixing was investigated for a range of operating conditions for both a Newtonian fluid (water) and softwood kraft pulp suspensions. The results show significant dependency of the jet mixing behavior on the fiber concentration. For water in the turbulent flow regime, mixing quality significantly improved with increasing jet-to-pipe velocity ratio. In addition, at the same jet-to-pipe velocity ratio, the jet penetration and mixing quality were almost independent of the mainstream velocity for a given jet-to-pipe diameter ratio.

For pulp fiber suspensions, on the other hand, the flow regime, which varies with the mainstream velocity, plays an important role in determining the degree of mixing. At low mainstream velocities, plug flow was approached, and mixing was poor due to the strong fiber networks, preventing tracer penetration. At higher velocities, the plug disintegrated, and the mixing quality significantly improved when the flow reached the turbulent regime. At higher mass concentrations ($C_m \geq 2\%$), however, mixing was poor, even at high mainstream velocities, since the shear was not high enough to disrupt the plug, so that additional shear was required to improve the mixing.

Nomenclature

C_m	suspension mass concentration or consistency [%]
D_j	jet-to-pipe diameter ratio
M'	modified mixing index [%]
R	jet-to-pipe velocity ratio
Re_j	jet Reynolds number
Re_p	pipe Reynolds number
U_p	main stream or pipe velocity [m/s]
U_j	side stream or jet velocity [m/s]
x/D	dimensionless distance downstream

References

Luettgen, C.O., J.D. Lindsay, and R.A. Stratton (1991), "Turbulent Dispersion in Pulp Flow: Preliminary Results and Implications for the Mechanisms of Fiber-Turbulence Interactions," *IPST Tech. Paper Ser.*, 408, November, pp. 1-29.

Pan, G. and H. Meng (2001), "Experimental study of turbulent mixing in a tee mixer using PIV and PLIF," *AIChE J.*, 47(12), pp. 2653-2665.

Rewatkar, V.B., R.J. Kerekes, C.P.J. Bennington (2002), "Use of Temperature Profiling to Measure Mixing Quality in Pulp Bleaching Operations", *Pulp Paper Can.*, 103(7), pp. T173-T179.

Robertson A.A. and S.G. Mason (1957), "The Flow Characteristics of Dilute Fiber Suspensions," *Tappi*, 40(5), pp. 326-334.

Tosun, G. (1987), "A Study of Micromixing in Tee Mixers," *Ind. and Eng. Chem. Res.*, 26(6), pp. 1184-1193.

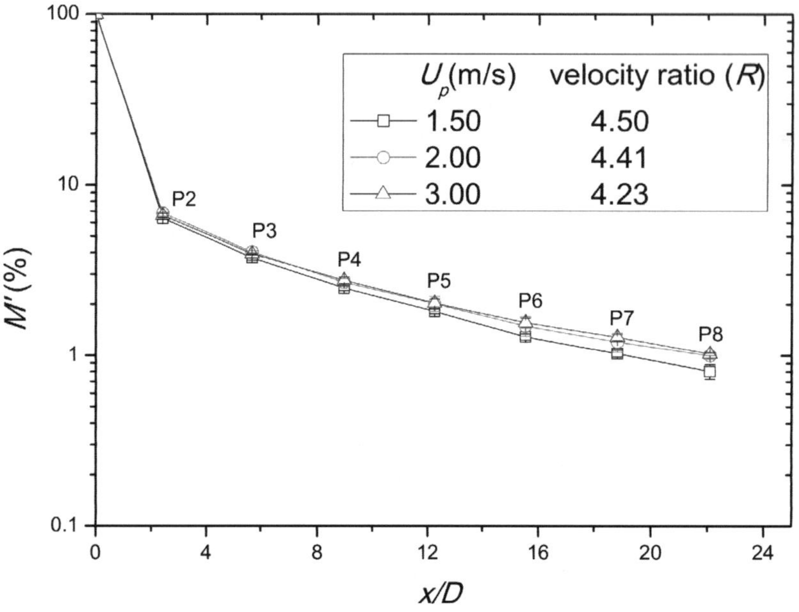

Figure 1. Modified mixing index as a function of dimensionless distance downstream, *x/D*, for Newtonian fluid (water) at U_p = 1.00 m/s, 2.00 m/s and 3.00 m/s for the same jet-to-pipe velocity ratio and D_r = 0.05.

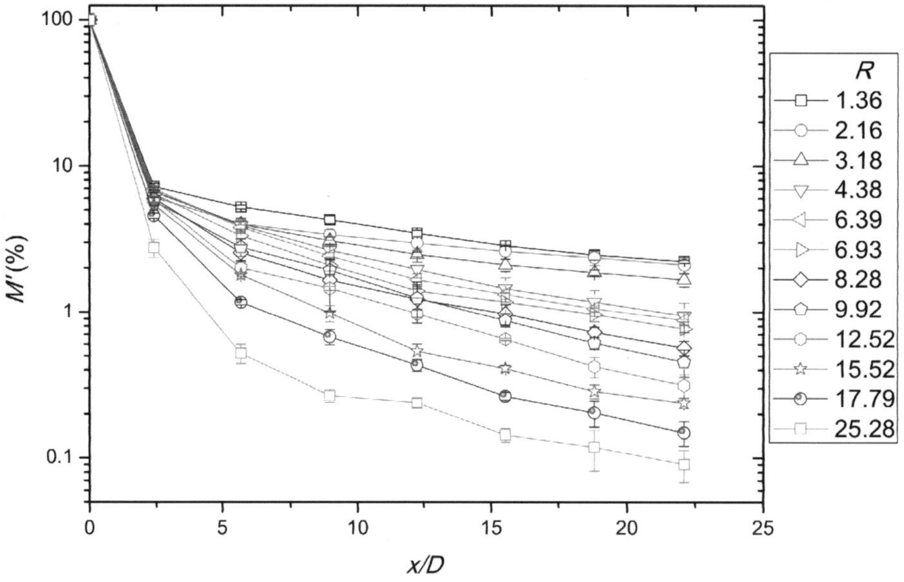

Figure 2. Modified mixing index as a function of *x/D* for various jet-to-pipe velocity ratios for Newtonian fluid (water) at D_r = 0.05.

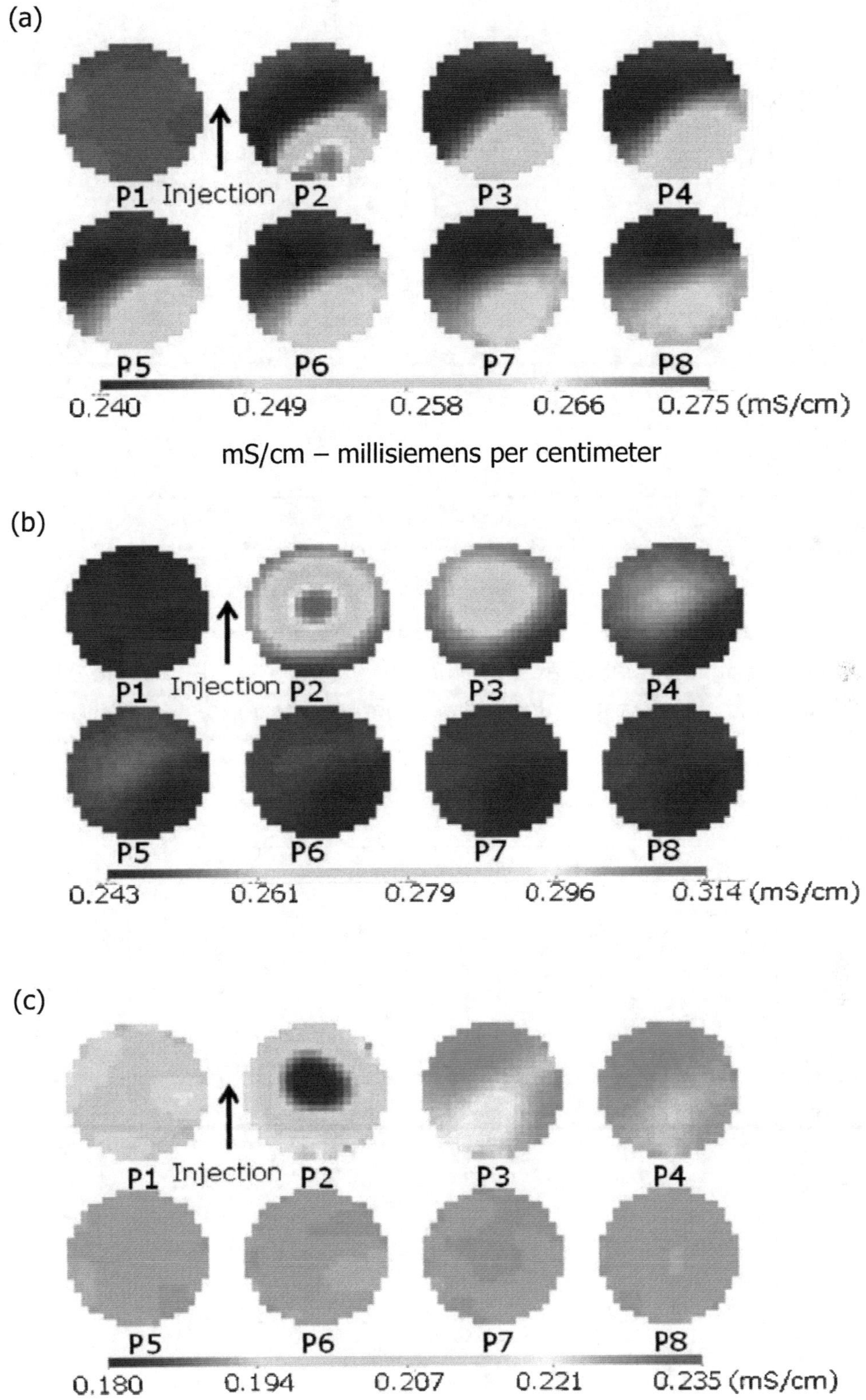

Figure 3. Tomographic images for in Newtonian fluid (water) for (a) wall-source at R = 2.22 (b) jet-mixing at R = 6.39 (c) jet-impaction at R = 15.74, D_r = 0.05.

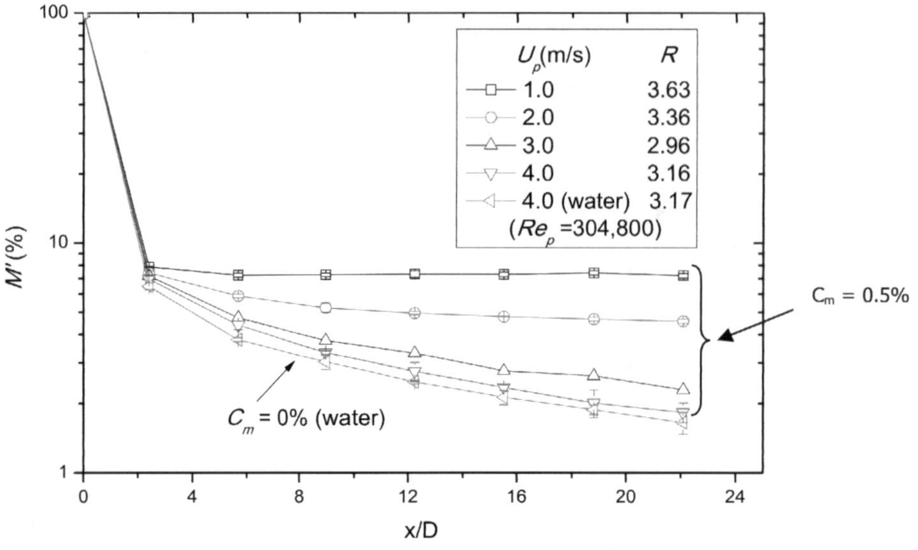

Figure 4. Modified mixing index as a function of *x/D* for C_m = 0.5% and 0%, D_r = 0.05 for various mainstream velocities and similar jet-to-pipe velocity ratios.

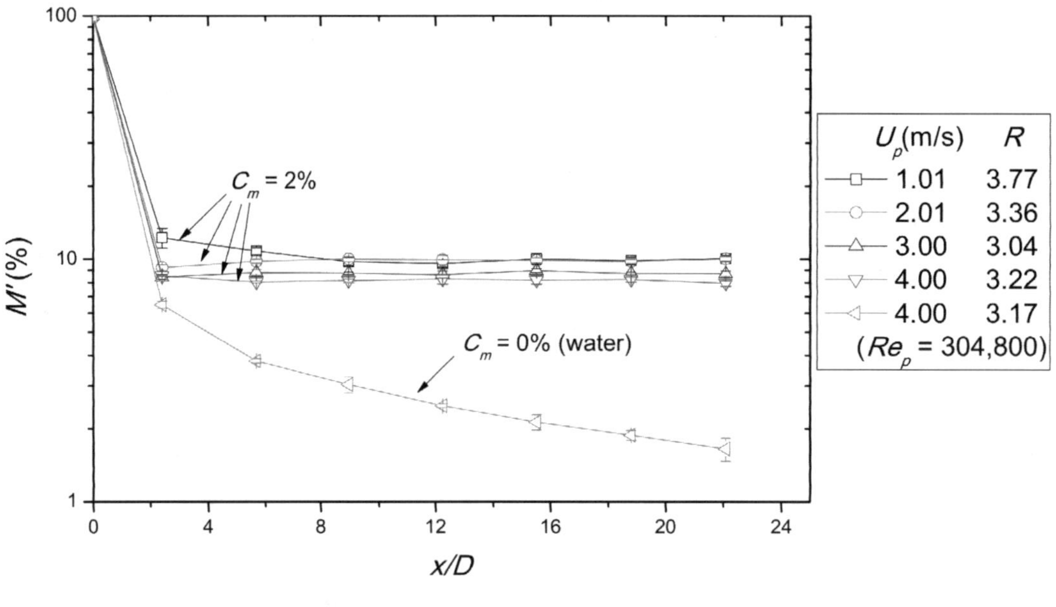

Figure 5. Modified mixing index as a function of *x/D* for C_m = 2.0% and 0%, D_r = 0.05 for various mainstream velocities and similar jet-to-pipe velocity ratios.

(36d) A Shear and Sedimentation Test Cell (SSTC) for Measuring the Effect of Mixing On the Performance of Chemical Additives

Marcio Bezerra Machado[1], Sujit Bhattacharya[2], Ng Samson[3] and Suzanne M. Kresta[1], (1)Chemical and Materials Engineering, University of Alberta, Edmonton, AB, Canada, (2)Syncrude Canada LTD., Edmonton, AB, Canada, (3)Syncrude Canada LTD., Edmonton, Canada

Mixing has an important role to play in the effectiveness of additive chemicals, but this interactions is poorly characterized in the literature. A uniform mixing environment is needed to test the effectiveness of demulsifiers and determine the mixing requirements of the process. To reach this objective, a mixing tank was developed: the SSTC (shear and sedimentation test cell). The SSTC is a narrow and tall tank (T=7.6 cm, H=3T) running with sets of multiple impellers. Three different impellers types were tested: A310, Rushton and Intermig. Sets of five (Rushton and A310) or six (Intermig) impellers of a relatively large diameter (D=T/2 for Rushton and A310 and D=2T/3 for Intermig) were used. Two fluids were used in the tank: water and triethylene glycol; to check the effect of the fluid viscosity. The power consumption of all sets of impellers was determined over a range of Reynolds numbers. Mean and rms velocities were measured throughout the tank using a Laser Doppler Velocimeter. Rushton impellers had the highest power consumption, followed by Intermig and A310. In all cases the set of impellers consumed less power than the same number of impellers operating separately. As the impeller to tank diameter ratio was large, the ratio between the maximum local dissipation and the bulk dissipation was smaller than the ratio obtained in a regular mixing tank. The maximum local energy dissipation was located in the impeller discharge stream for all configurations. The SSTC was able to provide a more uniform mixing for the bitumen extraction operation and is proposed as a test cell for additive chemicals in a wide range of applications.

See more of this Session: Mixing In Multi-Phase Systems I
See more of this Group/Topical: North American Mixing Forum

(36e) Measurement of Mixing Quality In Multiphase Systems

Jocelyn Doucet[1], Florian Bonniol[2], Francois Bertrand[2] and Jamal Chaouki[2], (1)Kengtek Inc., Longueuil, QC, Canada, (2)Department of Chemical Engineering, Ecole Polytechnique de Montreal, Montreal, QC, Canada

Commonly used methods to assess the state of a mixture are based on sampling and the measurement over time of the concentration of a key component at various locations, in order to determine how fast the variance of the sample concentration values (COV in fluid mixing and RSD in solids mixing) decreases to some asymptotic value. However, these tools are known to have intrinsic limitations due to their lack in capturing important spatial information. Moreover, their reliability is based on the number of samples considered, which can be cumbersome to obtain, be it experimentally or even numerically by post-processing simulation results.

The purpose of this talk is to show the potential of two mixing indices that were developed recently by our group to assess the mixing efficiency of fluid, granular and multiphase systems. Based on the concept of mixing in the weak and strong sense, they can be calculated from tracer trajectories either numerically through computer simulation or experimentally by means of non intrusive methods. We will recall how these two indices can provide the main mixing directions in a flow system as well as upper bounds for the mixing times. Three different cases will be discussed to illustrate the limitations of conventional mixing measures in properly characterizing mixing, and the wealth of detail provided by the two new indices.

See more of this Session: Mixing In Multi-Phase Systems I
See more of this Group/Topical: North American Mixing Forum

(36f) A Design Rule for Prediction of the Just Suspended SPEED of Mixed Slurries

Inci Ayranci[1], Theodore Ng[1], Maria Garcia[2], Arthur W. Etchells III[3] and Suzanne Kresta[1], (1)Chemical and Materials Engineering, University of Alberta, Edmonton, AB, Canada, (2)Chemical Engineering, Rowan University, Glassboro, NJ, (3)AWE3 Enterprises, Philadelphia, PA

ADESIGN RULE FOR PREDICTION OF THE JUST SUSPENDED SPEED OF MIXED SLURRIES

InciAyranci, Theodore Ng, Maria Garcia, Arthur W. Etchells, Suzanne M. Kresta

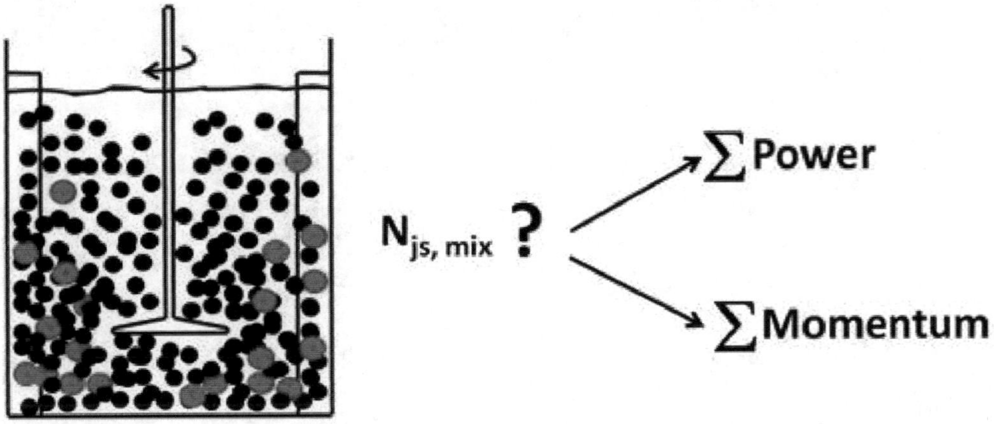

Designfor solids suspension in stirred tanks is based on an empirical correlationknown as the Zwietering correlation (Zwietering, 1958). This correlation hasmany limitations, one of which limits the industrial use significantly the correlation'sapplicability only for unimodal slurries. With the current design rule the justsuspended speed, N_{js}, of the mixed slurry is determined assumingthat the slurry consists of only the particle with the maximum N_{js}. Ourprevious work on binary slurries (Ayranci and Kresta, 2011) showed that thisdesign rule fails depending on the particle size, particle density and theconcentration of the two solid phases. The limitation of the design correlationfor mixed slurries and the failure of the current design rule clearly show thatthere is need for a model to predict the N_{js} of mixed slurries. The objectiveof this study is to propose a model to predict the N_{js} of mixed slurriesat relatively low concentrations where particle-particle interactions do notplay a significant role. We propose two models: the power model and the momentummodel. The power model states that the power required for the complete offbottom suspension of the mixture is a summation of the power of the unimodalslurries of the solid phases present in the mixture. Similarly, the momentummodel is based on the sum of momentum of the unimodal slurries. The two modelswere tested with binary mixtures of seven type of solids in water: nickel,bronze, glass of two particle sizes, sand, urea formaldehyde and ion exchangeresin. The total solids loadings were increased up to 35 wt% (weight percent). First,the unimodal slurry data was used to test the concentration dependence of theZwietering correlation. Then the two models were tested. The unimodal datashowed that the Zwietering correlation cannot predict the concentrationdependence accurately. The power model was successful in predicting the mixedslurry N_{js} up to 20 wt% solids loading.

See more of this Session: Mixing In Multi-Phase Systems I
See more of this Group/Topical: North American Mixing Forum

(102a) Determination of Drop Size Distribution In a Centrifugal Contactor

Nicholas B. Wyatt[1], **Timothy J. O'Hern**[2] and Bion Shelden[1], (1)Sandia National Laboratories, Albuquerque, NM, (2)Engineering Sciences, Sandia National Laboratories, Albuquerque, NM

Centrifugal contactors are widely used in separation and mass transfer applications involving immiscible phases. Quantifying drop size distributions in such processes is critical to understanding and tuning mass transfer characteristics between immiscible phases. Many studies have focused on the complex flows in both the annular and separation regions of the contactor. However, the magnitude and distribution of droplet sizes in the mixing region has not been characterized. For mass transfer applications, drop size and, thus, surface area are key factors. Through the use of high speed imaging, we measure drop size distributions in the annular mixing region for mixtures of silicone oil and water. We measure the evolution of the drop size distribution as the material travels from the mixing region to the vane region. Further, we quantify size distributions for air bubbles that are entrained in the flow. The distributions are examined over a range of rotor speeds from 1100 to 2100 rpm for silicone oil viscosities from 5 to 20 cSt. Drop sizes for both air and oil (dispersed phase) are found to be normally distributed at locations near the surface of the mixing zone. Near the bottom of the mixing zone, the distributions become log-normal and the mean drop diameter shifts to lower values.

Sandia National Laboratories is a multi-program laboratory operated by Sandia Corporation, a wholly owned subsidiary of Lockheed Martin Corporation, for the U.S. Department of Energy's National Nuclear Security Administration under contract DE-AC04-94AL85000.

See more of this Session: Mixing In Multi-Phase Systems II
See more of this Group/Topical: North American Mixing Forum

(102b) Effect of Impeller Submergence On Flow Characteristics and Minimum Agitation Speed for Solids Suspension In Partially Filled Stirred Vessels for Different Impeller off-Bottom Clearances and Impellers Sizes

Shilan Motamedvaziri and Piero Armenante, Otto H. York Department of Chemical, Biological and Pharmaceutical Engineering, New Jersey Institute of Technology, Newark, NJ

In many industrial applications, mixing vessels have liquid height-to-tank diameter ratio H/T equal to, or larger than, 1. However, there are many instances where this ratio is lower than 1, as in all those cases in which the vessel is emptied. Even when H/T<1, sufficient agitation must still be provided in order to attain the desired process objectives. When the impeller submergence is reduced as a result of lowering the liquid level, the fluid dynamics of even a single-phase stirred liquid can become quite complex, with different regimes possibly existing depending on the geometric characteristics of the system (such as impeller clearance, liquid height, or liquid head above the impeller). The objective of this work is to determine the minimum liquid levels and the critical impeller submergence for different impeller off-bottom distances (C_T/T=0.12, 0.18, 0.37, 0.54 and 0.68) and different impeller diameters (D/T=0.26, 0.31 and 0.42) where adequate mixing process can still be achieved, both in a single liquid phase and in solid-liquid suspensions.

Computational Fluid Dynamic (CFD) simulations as well as Particle Image Velocimetry (PIV) were used here to study the system's hydrodynamics in flat-bottom baffled vessels equipped with a single Disk Turbine (DT) for different height-to-tank diameter ratios when the liquid level is smaller than the tank diameter. A commercial mesh generator (Gambit 2.4.6) coupled with a CFD software package (Fluent 6.3.26) was used to simulate the flow field. A standard κ-ε model as well as a realizable κ-ε model coupled with enhanced wall treatment was used to model the turbulence flow. Results were obtained using a multiple references of frames (MRF) approach. For the case where air entrainment occurred, a Volume of Fluid (VOF) model coupled with MRF simulation was used. Experimental work was also carried out with a single liquid phase as well as with solid-liquid suspensions with 0.5 wt% dispersed solids. The minimum agitation speed for complete solid suspension, N_{js}, at different filling ratios was obtained visually and then they compared with CFD predictions.

In general, good agreement between the experimental data and the predicted results for the velocities distribution, pumping number and N_{js} were obtained. Both the experimental and the computational results, for different impeller off bottom clearances and impellers sizes, show that there is a minimum liquid level (or minimum impeller submergence) below which: (1) the macroscopic flow pattern changes substantially; (2) the pumping number drop significantly; (3) air entrainment occurs. When this occurred, it was observed that the average velocity field and turbulence intensity close to the tank bottom decreased substantially, which could be the reason why solid suspension becomes problematic.

See more of this Session: Mixing In Multi-Phase Systems II
See more of this Group/Topical: North American Mixing Forum

(102c) Aggregation In Homogeneous, Isotropic Turbulence

Jos Derksen, Department of Chemical & Materials Engineering, University of Alberta, Edmonton, AB, Canada

Aggregation in homogeneous, isotropic turbulence

Aggregation of solid particles suspended in a carrier fluidis an important particle growth mechanism. In many cases it is an unwantedphenomenon as it interferes negatively with processes designed for generatingparticle size distributions based on nucleation and growth. In other casesaggregation is desired as it for instance improves the filterability of a slurry. Aggregation and flow dynamics are tightlyconnected. For aggregation events to occur, particles need to collide and sticktogether. For the bigger particles we are interested in, collisions are mainlybrought about by velocity gradients in the flow field the particles aresuspended in (orthokinetic aggregation); not so muchby Brownian motion. The same velocity gradients that bring together particlesalso exert mechanical forces on aggregates that could break them again. Westudy this dynamic process of the generation of aggregate size distributions(ASD's) by means of numerical simulations that fully resolve the fluid andparticle dynamics. Turbulence with well-defined properties is created by meansof linear forcing in a fully periodic, three dimensional domain. In the domainwe release uniformly sized spherical primary particles that we make sticky byletting them interact through a square-well potential. We study the ASD and itsdynamics as a function of the strength of the turbulence, and as a function ofthe strength of the square-well potential. The figure shows some impressions ofthe simulations. In the left panel all particles in the domain are shown; theyare colored by the size of the aggregate they are part of (red: small; blue:big). The right panel shows the four biggest aggregate at the same moment intime (each aggregate has a different color; the periodic conditions show in thesense that the red aggregate is connected through the side boundaries).

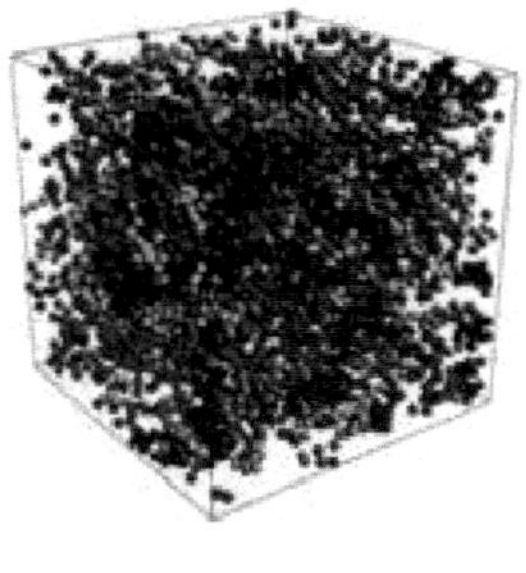

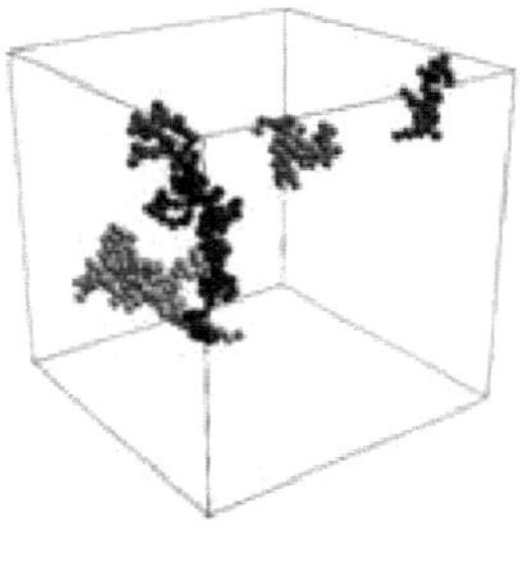

See more of this Session: Mixing In Multi-Phase Systems II
See more of this Group/Topical: North American Mixing Forum

(102d) Towards the Modelling of the Stabilization Process In Pickering Emulsions

Louis Fradette, URPEI, Chemical Engineering Department, Ecole Polytechnique de Montreal, Montreal, QC, Canada and Emir Tsabet, URPEI, Chemical Engineering, Ecole Polytechnique de Montreal, Montreal, QC, Canada

The replacement of traditional molecular surfactants by nano- or micrometer-sized solid particles in emulsification processes offers many advantages and creates new opportunities in the process industry. The emulsion stabilization mechanism with solid particles can be visualized in three steps: 1) the particle first approaches and reaches the fluid/fluid interface; 2) the particle adsorbs and gets trapped at the interface; 3) the adsorbed particles form a network that stabilizes the emulsion. At the moment, the state of knowledge does not allow for process design at the industrial scale mainly because of the lack of information about the overall stabilization mechanism. Being interested in emulsification process design, we are developing a methodology to predict the solid stabilized emulsions properties from the process conditions. The present work aims at developing a model describing the droplets stabilization on the basis of particle adsorption forces measurements using Atomic Force Microscopy. This approach allows us to identify the main parameters affecting the particles adsorption mechanism and therefore the stabilization rate, a condition to predict the droplets distribution size during the emulsification process.

See more of this Session: Mixing In Multi-Phase Systems II
See more of this Group/Topical: North American Mixing Forum

(102e) Effect of Oxygen Bubble Distributions On Rare Earth Precipitates In Molten Salts

Ryan W. Bezzant, Nuclear Engineering, University of Idaho-Idaho Falls, Idaho Falls, ID, Supathorn Phongikaroon, Department of Chemical Engineering and Nuclear Engineering Program , University of Idaho, Idaho Fall, ID and Michael F. Simpson, Pyroprocessing Technology Department, Idaho National Laboratory, Idaho Falls, ID

Effect of Oxygen BubbleDistributions on Rare Earth Precipitates in Molten Salts

Ryan W. Bezzant and Supathorn Phongikaroon

Universityof Idaho-Idaho Falls

Center forAdvanced Energy Studies

1776 Science Center Dr.

IdahoFalls, ID 83402 Michael F. Simpson Pyroprocessing Technology Department Idaho National Laboratory Idaho Falls, ID 83402

Abstract:

Pyroprocessing technologyhas been developed at the Idaho National Laboratory for treating spent nuclearfuel from Experimental Breeder Reactor-II. An electrorefiner (ER) is the keyunit operation of this process, recovering uranium metal via anodic dissolutionand electrodeposition at a cathode. During this process, minor actinides,fission products and rare earths (RE) form chlorides and accumulate in the LiCl-KCleutectic salt. This results in excess heat generation and other effects whichlower the system efficiency. To maintain ER performance, it is necessary toremove contaminated salt and refill with fresh salt. But this approach iscostly and generates a high waste volume.

Researchers at the KoreaAtomic Energy Research Institute have demonstrated oxygen sparging as aneffective way of separating rare earth chlorides from molten salt to partiallyprepare it for reuse. Rare earth chlorides react with oxygen at hightemperatures. Rare earth oxides and oxychlorides form and precipitate to thebottom of the salt container. After the salt cools, the salt can bemechanically separated from the rare earth fission products. Aside from ademonstration of the process, no fundamental aspects of this system are well understood.

Thus, to gain insight intophysiochemical parameters for systematic scale-up, such as mass transfer rateand gas holdup fraction, a transparent furnace with an oxygen sparging systemwas constructed with the capability of operating under an argon environment at500 °C. LiCl-KCl-$RECl_3$ternary salt was loaded in the quartz crucible to studythe effect of oxygen bubble sizes on rare earth precipitate in the system. Asimple two blade, downward pitch turbine was used for mixing the bubbles in themedium with a rotational speed varying between 100 and 250 RPM. The oxygenflowrate was varied between 0.05 L/min and 0.25 L/min. Oxygen concentrationwas detected by an oxygen sensor. Bubble sizes were captured by using a highspeed digital camera. Detailed results will be presented and discussed.

See more of this Session: Mixing In Multi-Phase Systems II
See more of this Group/Topical: North American Mixing Forum

(102f) Power Dissipation and Impeller Power Number In USP Dissolution Testing Apparatus 2

Yimin Wang and **Piero M. Armenante**, Otto H. York Department of Chemical, Biological and Pharmaceutical Engineering, New Jersey Institute of Technology, Newark, NJ

Dissolution testing is routinely used in the pharmaceutical industry to provide critical *in vitro* drug release information for quality control purposes and especially to assess batch-to-batch consistency of solid oral dosage forms such as tablets. The United States Pharmacopoeia (USP) lists several types of dissolution apparatuses. Of these, the most commonly used for solid dosage forms is the UPS Dissolution Testing Apparatus 2 consisting of an unbaffled, hemispherical-bottomed vessel equipped with a 2-blade radial impeller.

Significant information can be found in the literature of the power consumption of impeller-agitated, baffled systems. However, despite extensive industrial use and a large body of work on Apparatus 2, little or no information can be found in the literature on the power dissipated by this apparatus. Power consumption is a key parameter to predict for the rate of the dissolution process in such a system, especially for the prediction of mass transfer rate between the solid oral dosage form and the dissolution medium, since it has been shown previously that the power dissipated by an impeller rotating in a fluid correlates directly to the mass transfer coefficient experience by suspended particles. Thus, the determination of power dissipation is critical to quantify the mass transfer rate and hence be able to predict eventually the drug release rate from an oral dosage form during dissolution tests. Therefore, the main objective of this work was to experimentally determine and computationally predict the power dissipated by the impeller in the USP Dissolution Testing Apparatus 2 and the associated power number Po, and to relate these values to the mass transfer coefficient experienced by suspended drug particles.

A sensitive torque measurement apparatus was specifically built to measure the small torque applied to the liquid in the dissolution vessel when stirred by the paddle. To do so, the dissolution vessel and its content were floated in a shallow tank, and the force required to prevent rotation of this assembly was experimentally measured, from which torque and the power dissipation could be determined knowing the system geometry and the agitation speed.

A commercial CFD software package (FLUENT, 6.3.26; Gambit, 2.4.6) was used to predict the velocity distribution profiles in the USP 2 vessel, as well as the pressure distribution on the surface of the impeller blades from which the torque applied to the impeller and hence the impeller power dissipation were numerically calculated. Furthermore, the power dissipated by the impeller was additionally predicted by calculating, via CFD, the distribution of the local energy dissipation rate per unit mass in the USP 2 vessel, and then by integrating this variable over the entire liquid mass in the vessel.

The power dissipation was computationally and experimentally determined for a number of agitation speeds and liquid volumes. In general, good agreement was found between the experimental measurements and the computational results obtained with both methods. When Po was plotted against the impeller Reynolds number Re, a similar good agreement was typically found. As in other unbaffled system, Po was found to decrease slightly with Re, and was found to be in the range 0.25-0.38 for the case of the 900 mL volume fill level and in the range 0.20-0.29 for the 500-mL volume case. The values of the power dissipation per unit liquid mass were used to predict the mass transfer coefficient to suspended particles of different size and under different agitation conditions.

See more of this Session: Mixing In Multi-Phase Systems II
See more of this Group/Topical: North American Mixing Forum

(170a) 46 Years of Mixing Research Applied to Practical Applications

W. Roy Penney, University of Arkansas, Fayetteville, AR

Forty-six years of mixing research applied to practical application will be summarized. The research and applications include: Gas dispersion from vessel headspace; competitive fast reactions in agitated reactors, pump-around loops and pipeline mixers; backmixing in agitated compartmented columns, solids suspension at high solids loadings; vortex depth in partially-baffled vessels; just suspended speeds for liquid/liquid dispersions and mixing in high viscosity Vickers-Zimmer style polymerization reactors.

See more of this Session: Mixing Award & NAMF 21st Birthday Celebration
See more of this Group/Topical: North American Mixing Forum

(170d) NAMF 21st Birthday Celebration: Challenge Problems In Fluid Mixing Technology

Richard V. Calabrese, Chemical & Biomolecular Engineering, University of Maryland, College Park, MD and Fernando J. Muzzio, Chemical and Biochemical Engineering, Rutgers, the State University of New Jersey, Piscataway, NJ

Based on input from internationally recognized mixing experts, a set Challenge Problems in Fluid Mixing Technology will be presented that are relevant to national needs and targeted government funding. The purpose is to demonstrate the relevance of fluid mixing fundamentals over a broad spectrum of current application areas: and to provide researchers in academia, national laboratories and industrial consortia with a basis for attracting government and industrial funding.

See more of this Session: Mixing Award & NAMF 21st Birthday Celebration
See more of this Group/Topical: North American Mixing Forum

(239a) Effective Single-Impeller Turbulent Blending In Tall Vessels

Eric E. Janz[1], Kevin J. Myers[2], Jeremy K. Jones[2] and Julian B. Fasano[3], (1)Chemineer, Inc., Dayton, OH, (2)Chemical and Materials Engineering, University of Dayton, Dayton, OH, (3)Mixer Engineering Company, Troy, OH

Liquid level is typically reported to have a limited effect on blend time for an agitator with a single impeller in turbulent operation (blend time proportional to the square root of liquid level). However, this result seems to be based on limited data in shorter vessels with liquid levels less than or equal to the tank diameter. In practice, taller vessels are often used to increase volume without increasing footprint, and little information is available concerning the effect of higher liquid levels on agitator blend time.

This work demonstrates that turbulent blend times with single radial-flow and down-pumping axial-flow impellers increase rapidly as the liquid level becomes greater than the vessel diameter. One method to keep blend times from becoming excessive in tall vessels is to use multiple impellers. However, multiple radial-flow impellers "stage", with rapid blending occurring within the stage generated by each impeller but with significantly slower exchange between the stages. Multiple down-pumping axial-flow impellers do not stage such that they provide rapid blending, but this comes at the increased cost of additional impeller(s), larger shafting and gear box, and higher operating costs due to increased power draw.

A different approach to effective blending in tall vessels is to use a single up-pumping axial-flow impeller. By directing the impeller discharge flow into the majority of the liquid above the impeller, the velocities in the upper portion of the vessel are not as low as those produced by a down-pumping impeller such that blend time does not increase rapidly with increasing liquid level. Similarly, turbulent blend times do not increase rapidly with increasing liquid level when a single down-pumping impeller is placed high in the vessel such that the impeller's discharge flow is directed into the majority of the liquid below the impeller. Placing the down-pumping impeller high in the vessel mimics the advantageous velocity field generated by an up-pumping impeller placed low in the vessel by eliminating extremely low velocities.

This work presents data that can be used to design for effective turbulent blending using a single-impeller agitator across the wide range of liquid levels encountered in practice.

See more of this Session: Mixing In Single Phase Systems I
See more of this Group/Topical: North American Mixing Forum

(239b) Impeller Power Dissipation and Power Number for a Retreat-Blade Impeller Rotating In a Torispherical-Bottomed Tank Under Different Baffling Configurations

Nonjaros Chomcharn and **Piero M. Armenante**, Otto H. York Department of Chemical, Biological and Pharmaceutical Engineering, New Jersey Institute of Technology, Newark, NJ

Mechanically stirred glass-lined reactors and tanks are commonly used in the pharmaceutical and specialty chemicals industries, in order to avoid material compatibility issues between the vessel and its contents, thus minimizing possible product contamination problems. Because of the fabrication difficulties associated with glass lining metal surfaces, including baffles and impellers, these reactors are typically equipped with a single retreat curve impeller placed close to the bottom of the tank and a single baffle mounted from the top. Despite their common use in the industrial practice, these systems have not been extensively studied.

In this work, the power dissipation and the power number, Po, were experimentally obtained as a function of the impeller agitation speed in a 60-L laboratory plastic vessel having the same geometric shape, including a torispherical bottom, and the same geometric ratios of a typical industrial glass-lined reactor system. A typical scale-down version of single three-blade retreat-blade curved impeller was used in all experiments. Different baffling systems were studied, including a completely unbaffled tank, a partially baffled tank equipped with a single beaver-tail baffle commonly used in industry, and a fully baffled tank provided with four wall baffles. Water and aqueous sucrose solutions of different concentrations were used to vary the liquid viscosity and hence the Reynolds number. Torque was experimentally measured with a strain gage-based rotary torque transducer connected to an external multi-channel signal conditioner, display, and controller.

At lower Reynolds numbers (Re@100), all systems produced similar values of the power number, irrespective of baffling. However, as Re increased, the power dissipated by the baffled system increased appreciably above that of the partially baffled system and especially that for the unbaffled system. In addition, the power number decreased with increasing Re in all cases, but more significantly so for the unbaffled system. The value of Po approached an asymptotic value of about 0.7 for the baffled system. However, for the unbaffled system Po decreased with Re over the entire range of Reynolds numbers explored here (100<Re<300,000), finally approaching values for Po in the range 0.2-0.3 at large Re's. Po for the partially baffled case was intermediate between the fully baffled and unbaffled cases, with Po approaching 0.5, but was in general more similar in trend to the fully baffled system.

The Po-vs.-Re experimental data were fitted with a single type of regression curve containing three adjustable parameters, which were found to vary monotonically with increasing baffling effects. A good fit was obtained between the fitted curve and the experimental data. This equation can be used to predict the value of Po and hence the power dissipated in industrial systems.

See more of this Session: Mixing In Single Phase Systems I
See more of this Group/Topical: North American Mixing Forum

(239c) Effect of Operating Parameters On the Mixing Performance of the Superblend Coaxial Mixer

Louis Fradette[1], Xiao Wang[1], Katsuhide Takenaka[2] and Philippe Tanguy[3], (1)URPEI, Chemical Engineering Department, Ecole Polytechnique de Montreal, Montreal, QC, Canada, (2)R&D Dept., SHI Mechanical & Equipment Inc., Toyo City, Ehime, Japan, (3)Ecole Polytechnique de Montreal, Montreal, QC, Canada

The mixing performance of a Superblend coaxial mixer, which combines a Maxblend™ impeller as the central impeller and a double helical ribbon, was investigated experimentally in terms of power consumption and mixing time. The objective was to better understand the influence of the operating conditions on the mixing performance with Newtonian fluids. The experiment setup used allows each shaft to be driven independently. Taking advantage of individual torque-meter on each shaft and a well-developed discoloration technique, it was shown that the speed ratio $N_{Maxblend}/N_{ribbon}$(1, 2, 4, 6, 8) and the rotating mode (up- and down-pumping of the helical ribbon) have the largest influence on the mixing performance and that there is no universal operating mode for minimizing power and mixing time in all conditions and flow regimes. The results are presented in terms of the power consumption and mixing time throughout the laminar, transitional and turbulent regimes. The contribution of each impeller, the variation of flow patterns, the determination of the optimal operating conditions, and power constant are discussed.

See more of this Session: Mixing In Single Phase Systems I
See more of this Group/Topical: North American Mixing Forum

(239d) Investigation of the Dushman Reaction At Concentrations Relevant to Mixing Studies In Stirred Tank Reactors

Andreas Klbl[1], Vronique Desplantes[2], Laura Grundemann[3] and Stephan Scholl[3], (1)Institute for Micro Process Engineering (IMVT), Karlsruhe Institute of Technology, 76344 Eggenstein-Leopoldshafen, Germany, (2)Institute for Chemical and Thermal Process Engineering, TU Braunschweig, Braunschweig, Germany, (3)2Institute for Chemical and Thermal Process Engineering, TU Braunschweig, Braunschweig, Germany

Mixing processes both in stirred laboratory vessels and continuous flow mixers are frequently characterized with competing chemical reaction schemes. The most popular of these is the Iodide Iodate Reaction Method (also referred to as the Villermaux-Dushman method). Despite its popularity, the reaction kinetics of the Dushman reaction, which serves as a time scale for the examined mixing processes is poorly understood [Guichardon2000, Bourne2008]; it has only been investigated at concentrations far below those relevant to mixing studies. Experiments at concentrations relevant to mixing studies are rarely published [Kölbl2008, 2010]. We report on examinations in a 5L baffled laboratory tank equipped with a six-bladed Rushton turbine. Our aim is to provide more reliable data for the quantitative examination of mixing processes in stirred laboratory vessels at reactant concentrations relevant to mixing studies. Experimental results and investigation strategies are discussed.

Determination of the reaction order of the compounds

Through a systematic variation of the reactant concentrations, their relative influence on the velocity of the Dushman reaction can be examined. For continuous flow mixers the suggestions for the reaction orders of earlier examinations (H^+: 2; I^-: 2; IO_3^-: 1) [Guichardon2000] have been substantiated. For stirred laboratory tanks we suggest the following reaction orders: (H^+: 0.6; I^-: 1.5; IO_3^-: 0.6). The different reaction orders point to different reaction mechanisms effective during the investigation of continuous flow mixers and stirred laboratory vessels. Experimental results suggest that the reaction zone in a stirred laboratory tank is close to the injection needle, where high acid concentrations are found. The experimental characterization of continuous flow mixers are carried out at low acid concentrations. The difference in H^+ concentration might result in different reaction mechanisms.

Investigation of the influence of the acid employed

Experiments have been carried out with sulfuric acid (typically employed with the Iodide Iodate Reaction Method) and perchloric acid as a very strong acid. Opposed to previous findings [Kölbl2010], the choice of the acid in the case of stirred laboratory vessels has a considerable effect on the experimental results. Using perchloric acid, the triiodide concentrations obtained are 1.5 – 2 times higher than those obtained with sulfuric acid at the same experimental conditions. For the derivation of quantitative measures such as mixing times this experimental finding has to be considered.

Investigation of the influence of the ionic strength

Previous publications suggest a severe influence of the ionic strength on the reaction rate of the Dushman reaction [Guichardon2000]. We did not find any influence of the ionic strength.

We attribute the different findings to the use of potassium sulfate in the original examinations of a possible influence of the ionic strength. Sulfate acts a base especially in strong acid media, reducing the effective local H^+ concentration, thus resulting in reduced final triiodide concentrations.

[Bourne2008]

J.R. Bourne:

"Comments on the iodide/iodate method for characterising micromixing"

Chemical Engineering Journal 140(2008)638-641.

[Guichardon2000]

P. Guichardon, L. Falk, and J. Villermaux:

"Characterisation of micromixing efficiency by the iodide-iodate reaction system. Part II: kinetic study"

Chemical Engineering Science 55(2000)4245-4253.

[Kölbl2008]

A. Kölbl:

„Further Comments on the Iodide Iodate Reaction Method for Characterising Micromixing"

Chemical Engineering Journal 145(2008)176-177.

[Kölbl2010]

A. Kölbl and S. Schmidt-Lehr:

"The Iodide Iodate Reaction Method: The Choice of the Acid"

Chemical Engineering Science 65(2010)1897-1901.

See more of this Session: Mixing In Single Phase Systems I
See more of this Group/Topical: North American Mixing Forum

(239e) Investigation of Mixing In a Microscale Reactor Using Confocal μ -LIF

Yanxiang Shi, Department of Chemical and Biological Engineering, Iowa State University, Ames, IA, Michael Olsen, Mechanical Engineering, Iowa State University, Ames, IA and Rodney O. Fox, Department of Chemical & Biological Engineering, Iowa State University, Ames, IA

Mixing has a significant impact on many processes in chemical engineering and has been a bubbling area for decades. For example, 'Flash Nanoprecipitation', one promising technique that can be used to produce nanoparticles within small size distribution, requires fast mixing. To understand turbulent mixing in the reactors more thoroughly, the laser induced flourescence (LIF) method has been developed and applied by many researchers to large scale reactors. It is proved LIF is one of the best ways to measure the concentration fields. In this work, in conjunction with the confocal laser scanning microscopy, which is capable of improving the resolution, this method is applied to a microscale multi-inlet vortex (MIV) reactor.

Of the four inlet streams of the MIV, two opposing ones are dyed with Rhodamine 6G, in which way mixing is visualized. As in a previous study on MIV [1], mixing effects at three inlet Reynolds numbers, Re_j = 53, 93 and 240, are investigated, each representing a flow regime. Flow is steady and laminar at Re_j = 53 but is turbulent at Re_j = 240. Re_j = 93, however, falls into the transition range. As such, different mixing patterns between streams are detected at different Reynolds numbers. Due to scan speed limitation of the confocal system, line-scanning data at 21 different locations are acquired for statistical analysis. By subtracting the background noise from the mixing data and then normalizing it against the dye data, the mean concentration profiles and root mean square (RMS) of the fluctuations can be obtained. These experimental results are compared to large-eddy simulations to validate/improve the models. In addition to all that, a three dimensional reconstruction is also obtained to give us more intuitive insight on the mixing in the reactor.

[1] J. C. Cheng, M. G. Olsen and R. O. Fox. A microscale multi-inlet vortex nanoprecipitation reactor: Turbulence measurement and simulation. Applied Physics Letters. 2009(94):204104

See more of this Session: Mixing In Single Phase Systems I
See more of this Group/Topical: North American Mixing Forum

(239f) Mixing of Yield-Stress Fluids Using a Multilamination Mixer

Emilio J. Tozzi, Chemical Engineering and Materials Science, University of Calfiornia Davis, Davis, CA, Lori A. Bacca, Corporate Engineering Technologies Lab, The Procter and Gamble Co, West Chester, OH, William A. Hartt, The Procter and Gamble Co, West Chester, OH, Michael J McCarthy, Biological and Agricultural Engineering, University of Calfiornia Davis, Davis, CA and Kathryyn L McCarthy, Dept. Food Science and Technology, University of Calfiornia Davis, Davis, CA

Fluids that posess a yield-stress or are highly shear-thinning are difficult to mix in conventional agitated vessels due to the presence of unmixed low-speed zones far from the high-shear surfaces. Static mixers can be used to achieve mixing, though conventional ones have complex flow patterns that may complicate the experimental and theoretical assesment of mixing quality. A split-and-recombine (SAR) mixer on the other hand can generate a uniform pattern of parallel layers that are more easily characterized through optical or tomographic methods and are amenable to simplified theoretical treatments. A SAR mixer was constructed by stacking a series of plane plates with the desired flow pattern. The mixer was used to mix two carbopol gels of the same concentration and rheological properties, except that one contained a magnetic resonance contrast agent. Magnetic resonance imaging was used to determine the spatial distribution of the doped and undoped material at different stages of the static mixer. The images obtained show the feasibility of achieving a duplication in the number of layers upon exiting each stage. The three dimensional nature of the images obtained revealed detailed features of the fluid layers in planes normal or parallel to the flow direction. The effect of image resolution, imaging time and solute diffusion on image quality is discussed.

See more of this Session: Mixing In Single Phase Systems I
See more of this Group/Topical: North American Mixing Forum

(311a) Power Dissipation In a Partially Filled Stirred Vessel At Different Fill Ratios

Shilan Motamedvaziri, Sandra Lazo and Piero Armenante, Otto H. York Department of Chemical, Biological and Pharmaceutical Engineering, New Jersey Institute of Technology, Newark, NJ

The power dissipated by an agitated system is an important characteristic of any mixing process, since mechanical energy is required to homogenize the vessel content, disperse immiscible phases, suspend solids, increase mass transfer, and, in general, produce the desired mixing effect. Power dissipation in agitated systems depends not only on the type of impeller used, agitation speed, and the physical properties of fluid, but also on the geometric characteristics of the system, such as impeller clearance, liquid height, or liquid head above the impeller.

The main objective of this work was to investigate the effect of impeller submergence on power dissipation and Power Number, Po, in a flat-bottomed, baffled, cylindrical vessel stirred with a disk turbine (DT) for different impeller off-bottom clearances (C_T/T=0.12, 0.18, 0.37, 0.54 and 0.68) and different impeller diameters (D/T=0.26, 0.31 and 0.42) at constant impeller tip speed, U_{tip}. Additionally, the effect of impeller tip speed was assessed by changing the agitation speed for a fixed impeller size and position.

Experimental work was carried out with a single liquid phase. Torque was experimentally measured with a strain gage-based rotary torque transducer connected to an external multi-channel signal conditioner, display, and controller. Computational fluid dynamic (CFD) work was conducted here to predict the power number, velocity field, and turbulence intensity close to air-liquid interface for different height-to-tank diameter ratios when the liquid level was smaller than the tank diameter. A commercial mesh generator (Gambit 2.4.6) coupled with a CFD software package (Fluent 6.3.26) was used for this purpose. A standard κ-ε model as well as a realizable κ-ε model coupled with enhanced wall treatment was used to model the turbulence flow. Results were obtained using a multiple references of frames (MRF) approach. The velocity predictions were validated using Particle Image Velocimetry (PIV).

In general, good agreement between the experimental data and the predicted results for the power number and velocity distribution near the air-liquid interface were obtained. Both the experimental and the computational results for different impeller off-bottom clearances and impellers sizes show that there is a minimum liquid level (or minimum impeller submergence) below which: (1) the power number decrease significantly; (2) transition from "double-eight" to "single-eight" recirculation can be observed; (3) vortex formation and air entrainment occurs. When this happened, it was observed that the average velocity field and turbulence intensity close to the air-liquid interface increased substantially. This shift in flow pattern could be the reason for the air entrainment and the drop in power number.

(311c) Recent Advances In Mean Age Theory for Quantitative Mixing Analysis

Minye Liu, DuPont Engineering Technology, Wilmington, DE

Minye.liu@usa.dupont.com

In this talk, recent advances in the mean age theory will be discussed. This theory opens a whole new paradigm for quantitatively characterizing mixing processes in chemical reactors using CFD. It is a key extension to the current residence time distribution (RTD) theory. From this theory, mean residence time and higher moments of RTD can be computed with a small fraction of CPU time required by the current unsteady tracer tracking method but with significantly better accuracy[1]. From the solution of mean age, undesired features of a reactor, such as dead zones, short circuiting paths, etc., can easily be identified and characterized quantitatively. Such defects in reactor design can only be inferred or guessed using RTD theory.

With the solutions of mean age and the second moment of age, the degree of mixing can be computed for a steady continuous reactor[2]. Three variances are defined. The relations of these variances explain why RTD theory can only show the closeness of a mixing process to an ideal mixer but not quantify the mixing state. These variances and the degree of mixing can be used to quantitatively compare the effects of design and operating parameters of a continuous flow stirred tank reactor (CFSTR) such as reactor aspect ratio, impeller type and locations, and inlet/outlet sizes and locations, etc.[3] Mean age distribution can also be used to characterize the transient tracer concentration distribution in a CFSTR. The time history of tracer concentration inside the reactor can be divided into two stages, a location dependent initial stage and a location independent stationary stage. Time dependent tracer concentration equation needs to be solved for only the first stage which is about the batch blend time. The rest of the history can then be determined by the steady mean age solution. Mixing performance of a CFSTR can then be completely determined quantitatively. Mean age distribution also offers an extremely efficient method in calculating blend time in a batch stirred tank reactor. The predicted blend times are in excellent agreement with correlations in the literature[4]. The required computing resources are orders of magnitude smaller than the current method of tracking transient tracer concentration history. This method will enable CFD to become a reliable tool for industrial reactor design and scale-up.

References

[1] Liu, M. and Tilton, J.N., 2010, "Spatial Distributions of Mean Age and Higher Moments in Steady Continuous Flows," *AIChE J.*, **54**, 2561.

[2] Liu, M., 2011, "A method for computing the degree of mixing in steady continuous flow systems," *Chem. Eng. Sci.*, **66**, 3045.

[3] Liu, M., 2011, "Prediction of Tracer Concentration and Mixing in CFSTRs with Mean Age Distribution," *Ind. & Eng. Chem. Res.*, **50**, 5838.

[4] Liu, M., 2011, "Quantitative characterization of mixing in stirred tank reactors with mean age distribution," *Can. J. Chem. Eng.*, (available online).

See more of this Session: Mixing In Single Phase Systems II
See more of this Group/Topical: North American Mixing Forum

(311d) Stirred Tank Dynamic Mixing Analysis; Comparison of PIV-PLIF and LES

Michael D. Cloeter[1], Santhosh Ramalingam[1], Somnath Roy[2] and Sumanta Acharya[3], (1)Core R&D, Engineering Sciences, Fluid Mechanics & Mixing, The Dow Chemical Company, Freeport, TX, (2) Department of Mechanical Engineering, Indian Institute of Technology Patna, Bihar, India, (3) Department of Mechanical Engineering, Louisiana State University, Baton Rouge, LA

Stirred tanks are used throughout the chemical industry for many process operations such as blending, single- or multi-phase reactions, and crystallization, to name a few. They have been the topic of much study both in academia and industry. There is still much to be understood in terms of localized dynamic phenomena such as macro-instabilities. Dynamic modeling of single phase blending in stirred tanks is of interest as a necessary first step towards understanding more complex problems such as reacting and/ or multiphase flows.

To that end, turbulent blending in a stirred vessel was studied experimentally using simultaneous Particle Image Velocimetry (PIV) and Planar Laser-Induced Fluorescence (PLIF). The experimentation was compared to computational results from Large Eddy Simulation (LES). The configuration is shown in Figure 1. The vessel is a 17.5 inch inner diameter, dish-bottomed, fully baffled tank. A nine inch diameter 45 degree pitched blade turbine (PBT) was used as the impeller, rotating sequentially at 50, 100, and 150 rpm. The impeller Reynolds numbers were 44K, 88K, and 132K, respectively. The fluorescent dye was added thru 1/8 inch tubing at a velocity of 1.36 ft/s at a distance of 6 inches above the impeller tip and 45 degrees behind the laser sheet. PIV/PLIF image data were taken at two phase-locked positions, one with the impeller blade in the plane of the laser sheet, the other 45 degrees out of plane.

LES calculations were found to be in good agreement with the measurements, as shown in Figure 2. The general features of the flowfield are consistent with those of PBT with a downward directed impeller jet stream, formation of the trailing edge vortices and high levels of turbulent kinetic energy in the trailing vortex regions.

See more of this Session: Mixing In Single Phase Systems II
See more of this Group/Topical: North American Mixing Forum

(311e) Role of Streamwise Vortices On Mixing Process In Highly Excited Confined Mixing Layer

Zhao, W., University of South Carolina
Wang, G., University of South Carolina

The rapid mixing process in confined mixing layer in a pipe is investigated under different Reynolds numbers, velocity ratios and forcing intensities recently. The roles of spanwise and streamwise vortex structures are identified.

From the side view, it can be found the large scale spanwise vortex is intensively enhanced at some specific forcing frequencies, 3.5 Hz and 5.3 Hz respectively. However, the mixing effect at 3.5 Hz is obviously weaker than that at 5.3 Hz. The investigations indicate that, the peak magnitude of mean velocity of the secondary flow in vertical direction is much higher at 5.3 Hz, than at any other frequencies. The high velocity of the secondary flow leads to a large spreading rate of the mixing layer. The vertical mean velocity also stretches the spanwise vortices, and causes their earlier breakdown and the succedent rapid mixing.

The cause of the vertical mean velocity can be attributed to the important streamwise vortex structures. Experiments indicate that both the mean streamwise vorticity and fluctuated streamwise vorticity, are very sensitive to the forcing frequency of 5.3 Hz. At the frequency, the asymmetric counter-rotation vortex pairs are strongly intensified which generate vertical mean velocity. Meanwhile, the velocity profile is obviously distorted quasi periodically in spanwise. This further enhances the mixing process in spanwise direction. At the inlet section of mixing chamber, the streamwise vortices are dominant by corner vortices whose vorticity decays slowly along streamwise distance. However, after the region, the streamwise vorticity (estimated by enstrophy) increases with the streamwise position due to some unknown mechanisms. Further investigations will be conducted to study the mechanism.

See more Presentations from this session: Mixing In Single Phase Systems II

See more Sessions in this Group: North American Mixing Forum

(362a) Mass Transfer with Surface Chemical Reaction In Fischer-Tropsch Synthesis: Simulation by a Lattice Boltzmann Method

Mohammad R. Kamali[1], Jurriaan J.J. Gillissen[1], Sankaran Sundaresan[2] and Harry E.A. Van den Akker[1], (1)Dept. of Multi-Scale Physics, Delft University of Technology, Delft, Netherlands, (2) Department of Chemical Engineering, Princeton University, Princeton, NJ

Abstract

In the current study, the interaction of convection, diffusion and a surface reaction is simulated with the help of an advanced two-phase lattice Boltzmann method in the context of a Fischer-Tropsch synthesis. In our case of interest, carbon monoxide and hydrogen are converted into liquid hydrocarbons with the help of a catalyst deposited on the internal structures of a structured reactor. The liquid product and the heat of reaction are carried away from the reacting surface with the help of a fully wetting liquid. A dedicated FTS kinetic model was implemented on the reactive surfaces. Reaction rate and conversion are calculated in terms of the dimensionless Pclet and Damköhler numbers.

See more of this Session: Computational Fluid Dynamics In Chemical Reaction Engineering
See more of this Group/Topical: Catalysis and Reaction Engineering Division

(362b) Micro-Scale Modelling of Flow and Particle Transport In Porous Media Via CFD

Daniele Marchisio[1], Federica Lince[1], Tiziana Tosco[2] and Rajandrea Sethi[2], (1)Dismic, Politecnico di Torino, Torino, Italy, (2)Ditag, Politecnico di Torino, Torino, Italy

Porous media are ubiquitous in chemical engineering. They are involved in many separation processes (e.g., filtration, chromatography) and they play an important role in numerous chemically reacting systems such as in catalysis and in remediation of contaminated aquifers. Very often the fluids involved are characterized by very diverse rheological properties, ranging from classical Newtonian to non-Newtonian behaviors. Although the final applications are very different from each other and are typically characterized by very different features there are some common issues that despite the huge amount of experimental data still need to be addressed. These are generally related to the need of describing the porous medium at the macro-scale as a continuum defined in terms of a set of parameters that quantify its flow properties (e.g., porosity, permeability) and its particle transport characteristics (e.g., collection efficiency).

This work (co-funded by European Union project AQUAREHAB FP7 - Grant Agreement Nr. 226565) aims at tackling these issues from a fundamental perspective by the use of pore-scale (or micro-scale) simulations via computational fluid dynamics.

Porous media with different characteristics have been analyzed and used to extract some representative two-dimensional geometries of a few millimeters in size. Grids were created with Gambit whereas the flow of Newtonian and non-Newtonian (i.e., Carreau and Cross models) fluids through these media was investigated with Fluent 12.0. Data were analyzed in terms of the flow field and pressure drops across the media for different superficial velocities ranging from laminar to turbulent flows. Simulation predictions were then analyzed with different classical approaches (i.e., Darcy, Ergun, Darcy-Forchheimer equations) and the corresponding parameters were eventually extracted. Results show that the considered geometry are large enough to model and represent a fraction of the porous medium and allowed to correct and extend the Darcy-Forchheimer law to non-Newtonian fluids.

In the second part of the work the presence of solid particles suspended in the fluid was considered. The collection efficiency of the different porous media was then quantified with computational fluid dynamics by using an Eulerian approach. The gathered simulation predictions were then compared with classical approaches for the estimation of the collection efficiency of the porous medium (i.e., classical filtration theory). Results show that these theories generally under-predict the overall efficiency and a correction was eventually formulated. Future work includes the use and implementation of these corrections in macro-scale simulations.

ANSYS

Packing Low Tube-to-Particle Diameter Ratio Packed Beds for CFD Simulation

Genong Li,
Aniruddha Mukhopadhyay,
Yi Dai and Chi-Yang Cheng

Ansys Inc.
Sept. 2011

Introduction

ANSYS

- **Low D/d ratio packed beds ($D/d<5$) are widely used in fixed bed reactors.**

  - Porosity is highly non-uniform, flow channeling
  - Velocity, temperature, species are highly non-uniform

- **Limitations of the porous medium model**
  – Cannot capture many physics: local hot spots and runaway reactions
  – Characteristics of the packing: Porosity and S/V ratio profile
  – Effective viscosity, effective conductivity …

3D Fully Resolved Model

ANSYS

- **With the development of high performance computing and advances in CFD, it is possible to conduct fully resolved 3D simulations for packed beds especially at low D/d ratio**
  - Reasonable long reactor segments can be simulated within days time frame
- **Advantages of the fully resolved model**
  - It can provide detailed information of a packed bed reactor. Reactor design can rely more on simulation-based solution.
  - The information extracted from fully resolved simulations can be used to evaluate the performance of other reduced order methods, and provide input parameters for them.

Procedures and Challenges

ANSYS

- **Prepare packed bed geometry**
  - Development of a packing algorithm for random packing
- **Mesh the packed bed geometry**
  - A lot of point-to-point contacts, small gaps
- **Set up and run the model in a CFD solver**
- **Extract useful information**
  - Packing characteristics: porosity profile,...
  - Flow field information

Packed-bed Packing

- **Random packing algorithms fall under two main categories**
  - Dynamic techniques
    - It is usually conducted in a DEM code. The movement of every packing element is tracked till it is settled in the bed according to the Newton's law.
      - Considering more physics, relatively slow
  - Geometric techniques
    - Packing is generated based on purely geometric considerations. The method is very fast.
- **In this work, only cylindrical beds packed with spherical balls are considered. A geometric packing method is developed and used.**

Packing Algorithm

- We will put balls sequentially into the tube.
- The bottom layer of the tube is first filled.
- Given all the balls already in the tube, the next ball's sitting position is limited to a finite number of locations. We will search for all these possible positions.
- We will put the next ball at one of these possible positions (randomly or at the lowest possible position).
  - Note that the packing will be random as a result of this!
- We will continue this process until the whole cylindrical column is filled up.

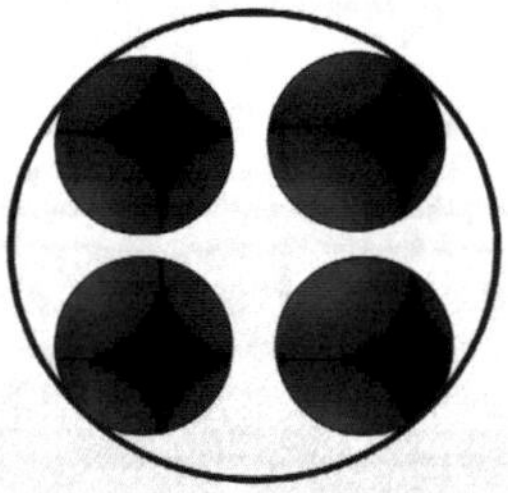

Illustration of the Packing Method

ANSYS

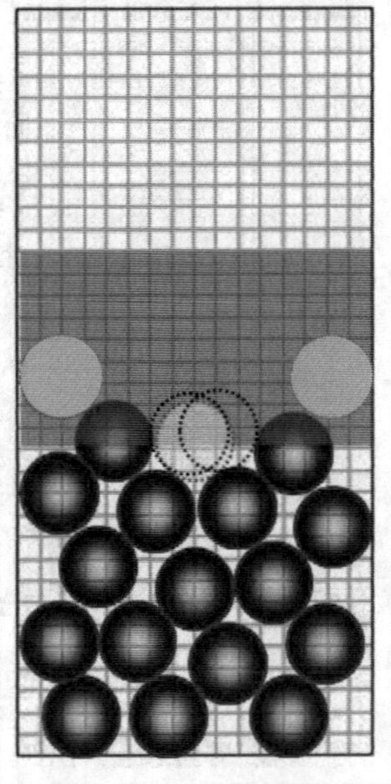

- A structured background mesh is created to facilitate the searching of the next ball's all possible positions.

$$\begin{cases} |\vec{x} - \vec{x_1}| &= r + r_1 \\ |\vec{x} - \vec{x_2}| &= r + r_2 \\ |\vec{x} - \vec{x_3}| &= r + r_3 \end{cases}$$

- Only the stable positions are considered.
- Packing options:
  - The lowest possible position
  - A random position
  - The combination of the two

Several Packing Examples

ANSYS

Mono-dispersed particles

Poly-dispersed particles

Meshing Problem and Solutions

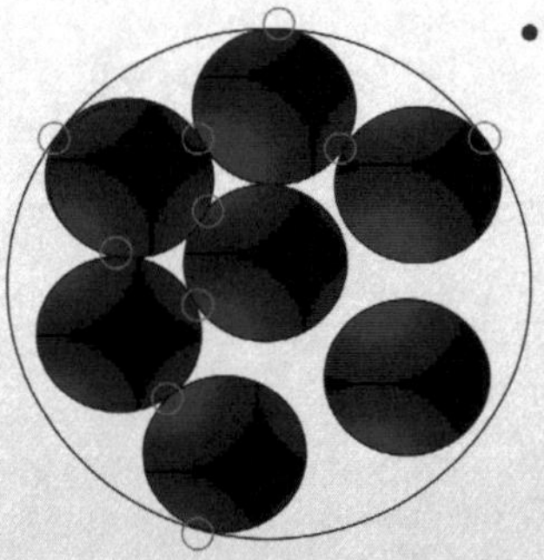

- A packed bed will have a lot of point-to-point contacts, which will cause problem in meshing!
- Solutions:
  - ➢ Method I: Packed balls are shrunk by a small percentage so there is always a gap between balls. Method is always robust, but there is no contact among balls. The heat conduction among balls cannot be considered. For the hydraulic simulation, the method should be fine.
  - ➢ Method II: Use "welding balls" to connect balls and ball and the wall.
- Size functions are used in meshing

Meshed Packed Bed

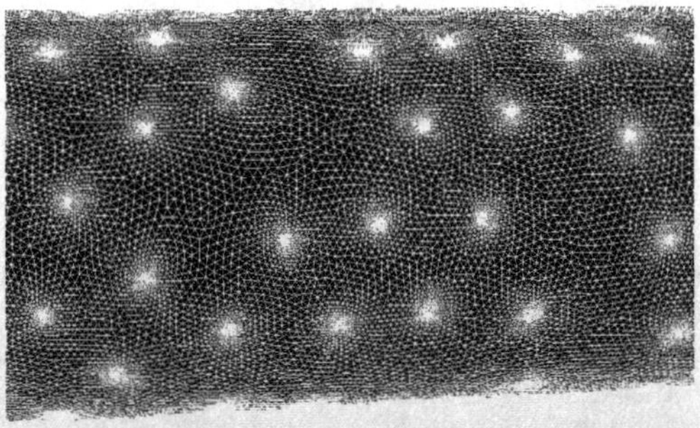

Using size function is critical to get high quality mesh with reasonable mesh count

Characteristics of Packing Beds

- **Once a packing structure is known, packing characteristics can be derived.**
- **Two packing characteristics are considered**
  - Porosity profile
    - Very important for flow distribution and pressure drop
  - S/V ratio
    - Very important for mass transfer and reaction.
- **A Monte Carlo method is developed to compute the above two parameters.**

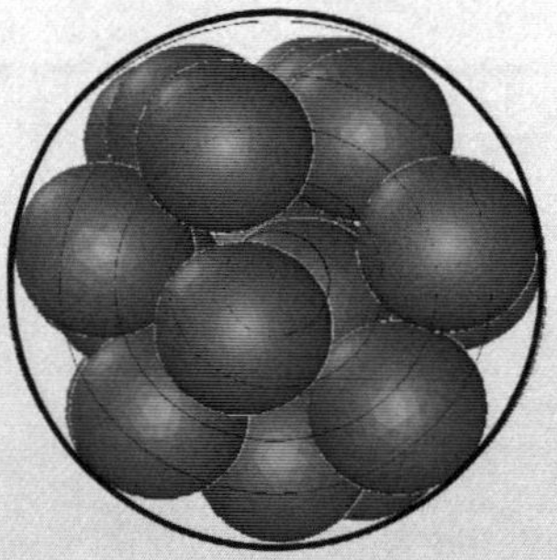

Packing Patterns at Different D/d Ratio

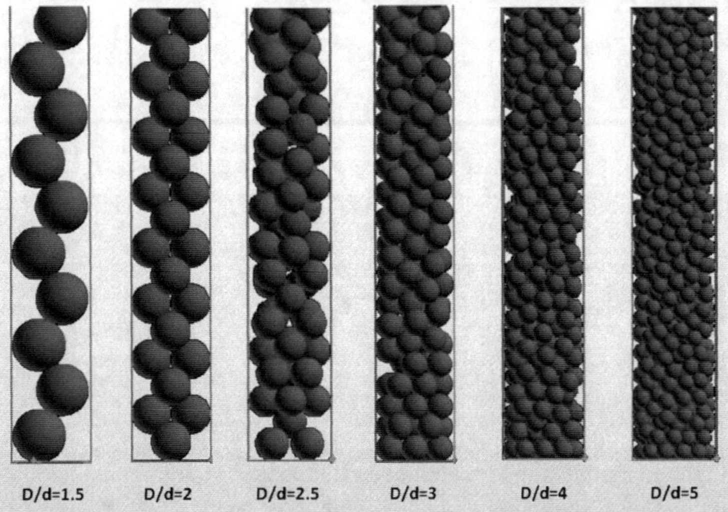

D/d≤2, structured packing; D/d>2, random packing

Porosity and S/V Ratio Profiles

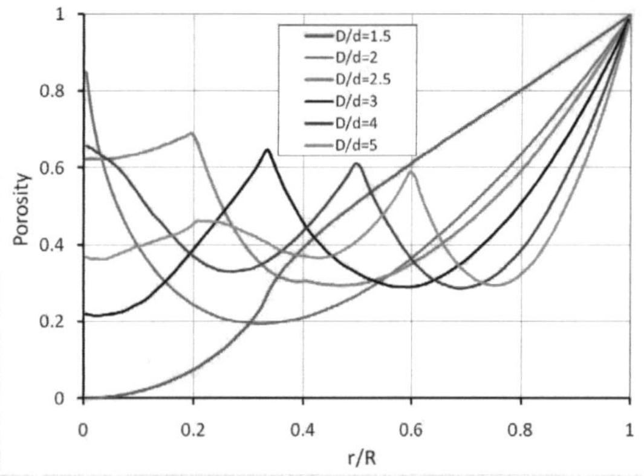

Porosity profiles for packed beds at different D/d ratios

Surface/volume ratio profiles for beds at different D/d ratios

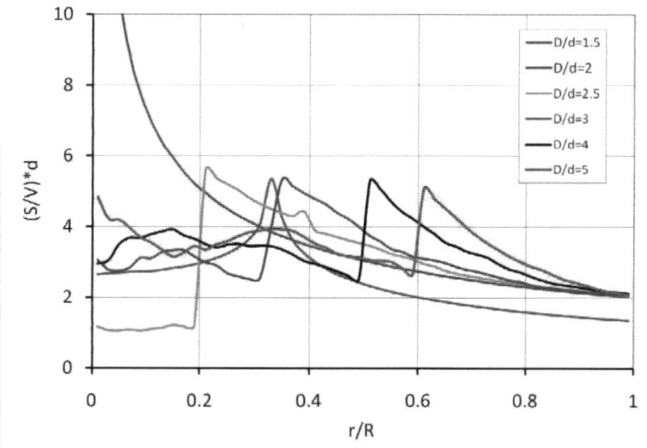

- The profiles highlight the non-uniformity of the packed beds.
- Such information can be used in the reduced order models!

Simulation of a Fully Resolved Bed

- **Dow chemical's plant scale reactor***
  - Tube diameter: D=26.12 mm
  - Tube length: H=2319 mm
  - Ball diameter: d=10 mm
- **Meshing detail**
  - Total number of balls: 916
  - Mesh size: 10M
- **Simulation condition**
  - Material: air
  - Mass flow rate: 3.65, 7.31, 10.96 kg/hr
  - Measured pressure drop: 884, 3491, 7593 Pa

* Bai et al., "A coupled DEM and CFD simulation of flow field and pressure drop in fixed bed reactor with randomly packed catalyst particles." Ind. Eng. Chem. Res. 2009;48:4060-4074.

Some Flow Results

ANSYS

Velocity: 0~20 m/s

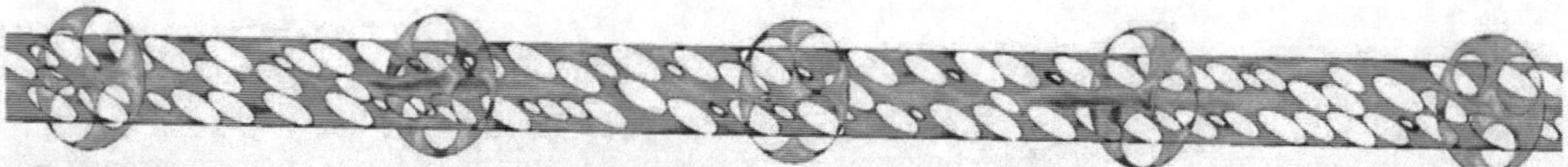

Pressure

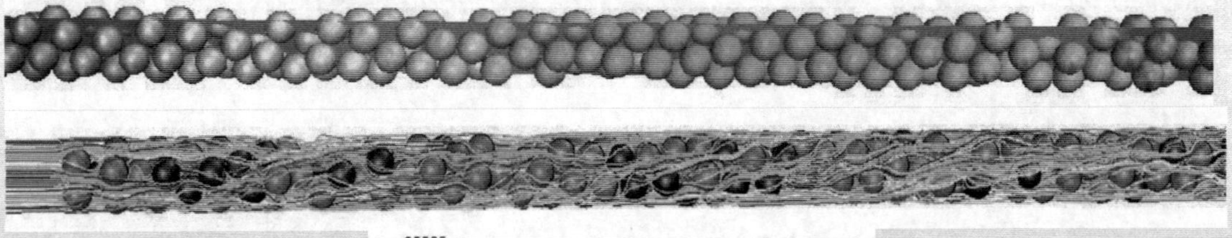

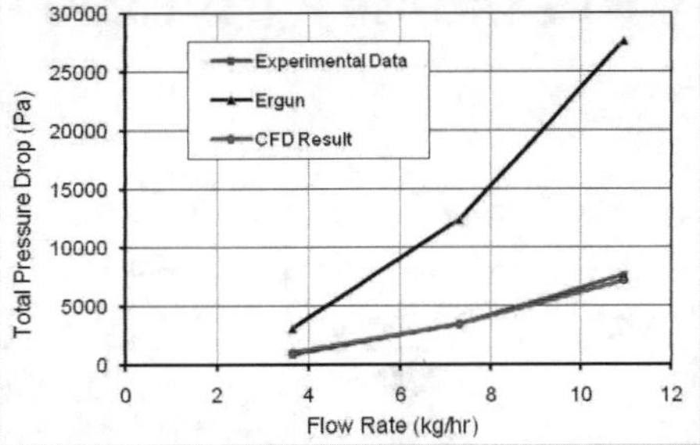

Averaged Velocity Profile

ANSYS

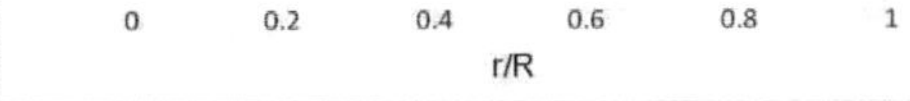

Porosity profile for Dow's plant-scale packed reactor

Averaged axial velocity along the reactor's radius

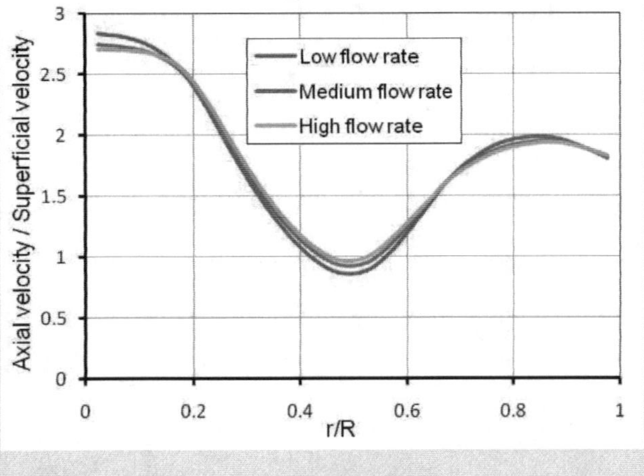

Such averaged profiles can be used to evaluate the accuracy of any reduced order model.

Conclusions

ANSYS

- A packing algorithm was developed to pack spherical balls into a cylindrical column randomly.
- Packing characteristics including porosity and surface/volume ratio were computed for low D/d ratio packed beds.
- Fully resolved simulations were conducted for a plant-scale real reactor. Results are in good agreement with experimental data.
- A fully resolved model provides not only a simulation-based solution in reactor design, but also a very useful tool to validate reduced order models.

(362d) CFD Simulations of An Ethylene Flame and Propylene Flares

Kanwar Devesh Singh[1], Hitesh Vaid[1], **Daniel H. Chen**[1], Helen Lou[1], Kuyen Li[1], Xianchang Li[2] and Christopher Martin[3], (1)Dan F. Smith Department of Chemical Engineering, Lamar University, Beaumont, TX, (2)Mechanical Engineering, Lamar University, Beaumont, TX, (3)Chemistry, Lamar University, Beaumont, TX

Air quality models (CAMx, CMAQ) often significantly under-predict observed peak O3 with the current VOC emission inventories in HGB/BPA. Even though this could be due to the inaccuracy of the photochemical models used, it is very likely that some VOC emissions may be under-reported or unreported due to the assumption of robust destruction efficiencies of industrial flares. From rigorous combustion chemistry, incomplete combustion of hydrocarbons is a likely source of free radicals such as perhydroxyl, radical producing species such as formaldehyde, and highly-reactive VOCs (HRVOCs) such as ethylene and propylene.

CHEMKIN CFD and FLUENT were used for our simulations. The simulation of laboratory McKenna burner is aimed at validating the turbulence model parameters and a 50-species reduced mechanism based on a combination of GRI 3.0 and USC mechanisms for the combustion of C1-C3 light hydrocarbons. The simulation of Tulsa flare tests is aimed at predicting the combustion efficiencies (CE and DRE) and hydrocarbon emissions based on validated CFD models. The simulations can establish the needed data base for finding the relationship between flare efficiencies/emissions and flare operating/ meteorological conditions.

See more of this Session: Computational Fluid Dynamics In Chemical Reaction Engineering
See more of this Group/Topical: Catalysis and Reaction Engineering Division

(362e) Comparative Study of Spouted Bed Hydrodynamics Based On Dimensionless Approach Using Optical Probes and CFD

Shreekanta Aradhya, Chemical and Biological Engineering, Missouri University of Science and Technology, Rolla, MO and Al-Dahhan Muthanna, Chemical & Biological Engineering, Missouri University of Science and Technology, Rolla, MO

@font-face { "Calibri"; }@font-face { "SimSun"; }p.MsoNormal, li.MsoNormal, div.MsoNormal { margin: 0in 0in 10pt; line-height: 115%; font-size: 11pt; ; }p.Paragraph, li.Paragraph, div.Paragraph { margin: 0in 0in 0.0001pt; text-indent: 0.25in; font-size: 10pt; ""; }span.go { }.MsoChpDefault { font-size: 10pt; ; }div.WordSection1 { page: WordSection1; }

Comparative Study of Spouted Bed Hydrodynamics Based on Dimensionless Approach using Optical Probes and CFD

[*]Shreekanta B Aradhya, [#]Xingying Lan and [*+]Muthanna Al-Dahhan

[*]Department of Chemical & Biological Engineering

[+]Department of Nuclear Engineering

Missouri University of Science & Technology, Rolla, MO – 65409, USA

[#]State Key Laboratory of Heavy Oil Processing, China University of Petroleum, P.R China - 102249

Email: sba2p4@mst.edu

Dimensional analysis has been used as a tool to develop a set of scaling parameters for proper design of fluidized beds. He et al. (1997) modified the scaling factors for fluidized beds proposed by Glicksman (1984) to provide a set of scaling parameters for the similarity behavior of spouted beds by adding the internal friction angle and the loose packed voidage. He et al. (1997) proposed the following controlling dimensionless parameters for a spouted bed: , , , , , sphericity of particles (), interfacial angle of particle (), loose peak voidage (), dimensionless particle size distribution, and dimensionless bed geometry. Using different spouted bed conditions that cover ambient and high temperature and pressure conditions and different particle types and sizes, He et al. (1997) demonstrated that close hydrodynamic similarity was maintained in the beds when these scaling parameters were closely matched.

However, the scale-up methodology of dimensional analysis has been evaluated only based on global parameters such as dimensionless spout diameters, dimensionless fountain height, and pressure along the bed height. No detailed hydrodynamics evaluation has been performed or reported. This work will focus on assessing the several match and mismatch conditions (based on dimensionless groups) reported by He et.al (1997) by using advanced optical probes and by CFD simulations. The parameters will be evaluated with the help of CFD simulations and optical probe measurements, which will be compared not only based on global parameters, but also local parameters such as solids holdup, solids velocity and their fluctuations.

See more of this Session: Computational Fluid Dynamics In Chemical Reaction Engineering
See more of this Group/Topical: Catalysis and Reaction Engineering Division

CFD STUDY OF HEAVY OIL GASIFICATION IN AN ENTRAINED-FLOW GASIFIER

Vikram Sreedharan, Dept. of Biotechnology, Chemistry & Environmental Engineering, Aalborg University, Denmark*
Bjørn H. Hjertager, Faculty of Science & Technology, University of Stavanger, Norway
Tron Solberg, Dept. of Biotechnology, Chemistry & Environmental Engineering, Aalborg University, Denmark

E-mail: vs@bio.aau.dk

Introduction

Entrained-flow gasifiers are widely used for the gasification of coal and petcoke (fluidized bed gasifiers being another popular choice), especially for the synthesis of chemicals, in power generation and for the production of synthetic fuels. This gasifier type has also been historically associated with the gasification of heavy oil, for instance in the Shell gasification process (SGP) and in Texaco gasification studies. Entrained-flow gasifiers are characterized by high combustion-zone temperatures (in excess of 1400°C) and an operating pressure between 20 and 70bar (Higman & Burgt, 2008). These high temperatures are necessary to ensure high carbon conversion despite the short residence time of the feedstock within the reactor.

With the aim of promoting fuel security, diversification and an increase in thermal efficiency, feeds such as Orimulsion™ (heavy oil) are being considered for use in combined-cycle power plants. Despite a number of studies focusing on the investigation of gasification of solid fuels, only a limited number of studies are found in the literature for the gasification of liquid fuels, most of which deal with the gasification of black liquor. The current study aims to model the physical and chemical processes taking place within the entrained-flow heavy oil gasifier using computational fluid dynamics (CFD). This model needs to account for a number of complex processes such as turbulence, multiphase flow, heat transfer, mass transfer and chemical reactions. In this study, the experimental results of Ashizawa et al (2005) have been compared with the predictions obtained from the CFD model.

The focus of this work is to describe accurately the different chemical processes occurring during gasification, i.e., devolatilization, char combustion and the gas-phase reactions for the heavy oil. Furthermore, turbulent multiphase (dilute) flow with combustion is modeled, with the use of an appropriate model for radiation.

Theoretical Background

When the feed is introduced into the reactor at high temperature, with a "carrier" medium such as air or oxygen, a number of chemical and physical processes occur within the gasifier. In order to obtain the benefit of savings in space and energy, gasification takes place at high pressures. If the feedstock used is heavy oil, the volatile content is consumed during the evaporation/devolatilization phase and the resulting char is burned or gasified.

In an entrained-flow gasifier, the path of the heavy oil droplets would be largely influenced by the gas flow which is represented typically as a dilute flow regime. In this regime, the volume occupied by the droplets and the droplet-droplet interactions are assumed to be negligible. A criterion used in the CFD software FLUENT for dilute flow is that the particle volume fraction should not exceed 10%.

Multiphase Flow

An Euler-Lagrange model known as the discrete phase model (DPM) is appropriate for multiphase flows where the dispersed phases are dilute. The continuous (gas) phase is described in an Eulerian formulation by solving conservation equations for mass, momentum and energy, while the dispersed phase (droplets) is described in a Lagrangian formulation by tracking the exchange of mass, momentum and energy for individual particles.

Continuous Phase

The reacting flow in the continuous gas phase is of a turbulent nature, and can be described by Favre-averaged balance equations for mass, momentum and energy. The governing transport equations can be written in a generalized form using Cartesian tensor notation as

$$\frac{\partial}{\partial t}(\rho\Phi)+\frac{\partial}{\partial x_j}(\rho u_j\Phi)=\frac{\partial}{\partial x_j}\left(\frac{\mu_t}{\sigma_\Phi}\frac{\partial\Phi}{\partial x_j}\right)+S_\Phi+S_{\Phi P} \tag{1}$$

In the above expression, the dependent variable $\Phi = 1, u_i, h, Y_k, k, \varepsilon$ representing variables such as mass, momentum, energy, species, turbulent kinetic energy and turbulent energy dissipation rate, respectively. S_Φ is a standard source term in the governing equation (1) and $S_{\Phi P}$ is a source term taking into account the interaction of the continuous (gas) phase with the dispersed particle (droplet) phase. The turbulent eddy viscosity is given by

$$\mu_t = \rho C_\mu \frac{k^2}{\varepsilon} \tag{2}$$

This eddy viscosity is determined from the Standard $k-\varepsilon$ turbulence model.

Discrete Phase

The conservation of mass equation for the discrete phase (oil droplets) takes the form

$$\frac{dm_p}{dt}=\frac{dm_1}{dt}+\frac{dm_2}{dt}+\frac{dm_3}{dt}+\frac{dm_4}{dt}+\frac{dm_5}{dt} \tag{3}$$

Where, $m_1 - m_5$ are the masses of gas originating from chemical reactions (11) – (15). The trajectory of the discrete phase droplets are determined from Newton's second law of motion. The force balance for a droplet can then be written as

$$\frac{du_{pi}}{dt}=F_D\left(u_{gi}-u_{pi}\right)+g_i\frac{\left(\rho_p-\rho_g\right)}{\rho_p}+\frac{\rho_g}{2\rho_p}\frac{d}{dt}\left(u_{gi}-u_{pi}\right)+\left(\frac{\rho_g}{\rho_p}\right)u_{pj}\frac{\partial u_{gi}}{\partial x_j} \tag{4}$$

Here, the subscripts g and p refer to the gas and particle (droplet) phase, respectively. The droplet inertia is balanced by the drag, the gravity, the virtual mass and the force due to the pressure gradient acting on the droplet. For a rigid spherical particle (droplet), the drag is defined by

$$F_D=\frac{18\mu_g}{\rho_p d_p^2}\frac{\mathrm{Re}\,C_D}{24} \tag{5}$$

The drag coefficient is determined from the Schiller and Naumann model correlation

$$C_D=\begin{cases}24\left(1+0.15\mathrm{Re}^{0.687}\right)/\mathrm{Re} & \mathrm{Re}\le 1000\\ 0.44 & \mathrm{Re}>1000\end{cases} \tag{6}$$

Thermal energy conservation for the discrete phase particle (droplet) can be written as

$$\frac{d\left(m_p c_p T_p\right)}{dt}=A_p\frac{\lambda Nu}{d_p}\left(T_g-T_p\right)+A_p\varepsilon_a\left(Q_R-\sigma_B T_p^4\right)+\sum_{k=1}^{5}\dot{m}_k\Delta h_k \tag{7}$$

In the above expression, the rate of change of thermal energy for a droplet is balanced by the convective and radiative heat transfer as well as the heat due to evaporation, devolatilization and particle surface chemical reactions. The convective heat transfer coefficient is calculated from the Ranz and Marshall's Nusselt correlation

$$Nu=\frac{hd_p}{k_\infty}=2.0+0.6\,\mathrm{Re}_d^{1/2}\,\mathrm{Pr}^{1/3} \tag{8}$$

Heavy Oil Gasification

The gasification of heavy oil can be broken down into the following simple sub-processes for ease of modeling.

Passive Heating

The heavy oil droplets are heated till the vaporization/devolatilization temperature is reached. There is no mass transfer or chemical reaction during this stage. When the moisture in the heavy oil reaches a certain temperature - say the vaporization temperature, it is released. Meanwhile, energy is taken out from the gas phase to supply the latent heat of vaporization.

Devolatilization

A phenomenological model has to be constructed based on data from experiments used to characterize the heavy oil. This model is then used to predict the yields of some major gas components while preserving a strict elemental balance to determine stoichiometry (Shi et al, 2006). The main species included in the devolatilization model are CH_4, CO_2, CO, O_2, H_2, H_2S, SO_2, N_2 and H_2O. The release of volatiles from the heavy oil can be represented as

$$C_{m01}H_{m02}O_{m03}N_{m04}S_{m05}Ash_{m06} \rightarrow C_{m11}H_{m12}O_{m13}N_{m14}S_{m15} + C_{m21}Ash_{m26} \quad (9)$$

The kinetic devolatilization rate is then determined from the Kobayashi model, which is defined by two competing rates. The weighted rate of devolatilization is given by

$$\frac{dm_v}{dt} = (m_{p,0} - m_a)(\alpha_1 R_1 + \alpha_2 R_2)\exp\left(-\int_0^t (R_1 + R_2)dt\right) \quad (10)$$

$R_1 = A_1 \exp(-E_1/RT_p)$ and $R_2 = A_2 \exp(-E_2/RT_p)$ are the competing rates that control devolatilization over different temperature ranges. The first yield factor α_1 is the fraction of volatiles as determined by proximate analysis of the heavy oil, representing devolatilization at low temperature. The second yield factor α_2 is taken as unity, which is the yield of volatiles at high temperature.

Oxidation of Volatiles, Char Combustion and Gasification

After the complete release of volatiles, its oxidation, char combustion and gasification take place until all the char is consumed or the particles flow out of the reactor. The chemical reactions include volatile oxidation, char combustion, char-steam gasification and char-carbon dioxide gasification, followed by the gas phase reactions. Volatiles oxidation/combustion for oxygen-rich conditions are described by,

$$C_{m11}H_{m12}O_{m13}N_{m14}S_{m15} + \left(m_{11} + \frac{m_{12}}{4} - \frac{m_{13}}{2} + m_{15}\right)O_2 \xrightarrow{k1} m_{11}CO_2 + \frac{m_{12}}{2}H_2O + \frac{m_{14}}{2}N_2 + m_{15}SO_2 \quad (11)$$

Using the expression for the partial oxidation of volatiles given in Chen et al (2000), the chemical reaction for volatiles oxidation/combustion for oxygen-lean conditions, that is, when $\phi < \left(m_{11} + \frac{m_{12}}{4} - \frac{m_{13}}{2} + m_{15}\right)$ is given by

$$C_{m11}H_{m12}O_{m13}N_{m14}S_{m15} + \phi O_2 \xrightarrow{k2} m_{11}(\alpha_1 CO_2 + (1-\alpha_1)CO) + \frac{(1-\alpha_1)}{2}(m_{12} - 2m_{15})H_2 + \frac{\alpha_1(m_{12} - 2m_{15})}{2}H_2O + \frac{m_{14}}{2}N_2 + m_{15}SO_2 \quad (12)$$

Table 1: Stoichiometric coefficient values

Parameter	Value
m11	14.88
m12	27.96

m13	0.072
m14	0.124
m15	0.336

The stoichiometric coefficient values in Table 1 are determined by the proximate and ultimate analysis of heavy oil (Orimulsion) given in Ashizawa et al (2005). If we substitute the parameter values from Table 1 in chemical equations (11) and (12), we get the complete and partial oxidation relations for the released volatiles for heavy oil. Carbon monoxide and hydrogen are produced during the gasification of the char particles contained in heavy oil. The performance of a gasifier is determined primarily by the char gasification reaction. The following heterogeneous (particle-surface) reactions have been included in the model

$$C + \left(1 - \frac{\gamma}{2}\right)O_2 \xrightarrow{k3} \gamma CO + (1-\gamma)CO_2 \tag{13}$$

$$C + CO_2 \xrightarrow{k4} 2CO \tag{14}$$

$$C + H_2O \xrightarrow{k5} CO + H_2 \tag{15}$$

The heterogeneous surface reaction rate is determined by a kinetics-diffusion limited scheme given by

$$\frac{dm_p}{dt} = -A_p p_{ox} \frac{D_0 R}{D_0 + R} \tag{16}$$

$$D_0 = C_1 \frac{\left[(T_p + T_\infty)/2\right]^{0.75}}{d_p} \tag{17}$$

and, the kinetic rate for the chemical reactions (13) – (15) is given by

$$R = C_2 e^{-(E/RT_p)} \tag{18}$$

Gas Phase Reactions

The continuous gas phase is modeled using a reduced global reaction scheme to describe gas phase chemistry. Chemical reaction kinetics appropriate for gasification studies have been used for the reaction scheme in this analysis. The following reaction paths have been considered for this model

$$CO + H_2O \underset{k6b}{\overset{k6f}{\rightleftarrows}} CO_2 + H_2 \tag{19}$$

$$CH_4 + H_2O \underset{k7b}{\overset{k7f}{\rightleftarrows}} CO + 3H_2 \tag{20}$$

The gas phase chemistry is modeled by combining laminar finite-rate chemical kinetics and eddy-dissipation turbulence-chemistry interaction. The values of the pre-exponential coefficient and the activation energies for the various chemical reactions considered for the gasification model (to determine the reaction rates) are given in Table 2.

Table 2: Chemical kinetics reaction rates

Eq. #	A_r	E_r (J/kmol)	n_r	$\prod_j [C_{j,r}]^{(\eta'_{j,r}+\eta''_{j,r})}$	Source
k1	2.119×10^{11}	2.027×10^{8}	0	$[Vol]^{0.2}[O_2]^{1.3}$	FLUENT user guide
k2	4.4×10^{11}	1.25×10^{8}	0	$[Vol]^{0.2}[O_2]^{1.3}$	Jones & Lindstedt (1988)
k3	1.36×10^{6}	1.3×10^{8}	0.68	-	Watanabe et al (2002)
k4	2.07×10^{7}	2.2×10^{8}	0.733	-	Watanabe et al (2002)
k5	2.07×10^{7}	2.2×10^{8}	0.733	-	Watanabe et al (2002)
k6f	2.5×10^{5}	1.38×10^{8}	0	$[CO]^1[H_2O]^1$	Maki & Miura (1997)
k6b	9.4×10^{6}	4.12×10^{8}	0	$[CO_2]^1[H_2]^1$	Maki & Miura (1997)
k7f	9.1×10^{7}	1.31×10^{8}	0	$[CH_4]^1[H_2O]^1$	Maki & Miura (1997)
k7b	5.5×10^{-6}	1.76×10^{9}	0	$[CO]^1[H_2]^3$	Maki & Miura (1997)

Model Description

An experimental study of a pilot-scale entrained-flow heavy oil gasifier by Ashizawa et al (2005) provides a framework of comparison for the current work. A simplified schematic of the heavy oil entrained-flow gasifier is depicted in Figure 1. The gasifier is cylindrical with a length of 3m and a diameter of 0.3m. The diameter of the inlet is 0.075m and that of the outlet is 0.15m. The computational mesh consists of 300 x 200 quadrilateral control volumes. The discrete phase method in ANSYS FLUENT 13.0 was used for the CFD simulation.

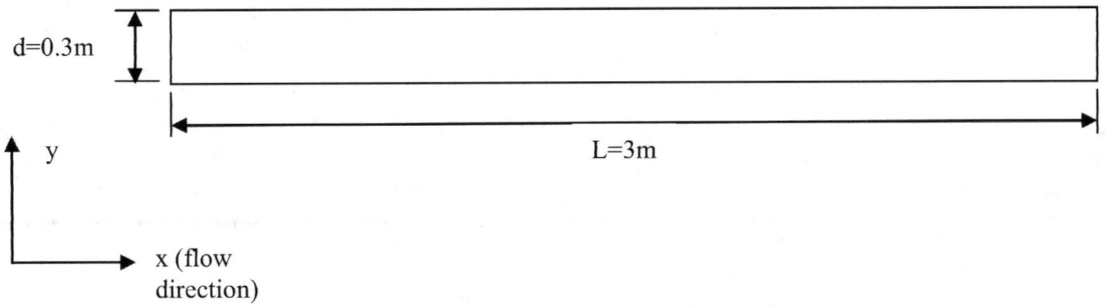

Figure 1: Gasifier schematic

Flow and Heat Transfer Models

The standard k-ε turbulence model was used for the continuous phase, and stochastic tracking was implemented in a steady-state model of the gasifier to simulate particle motion and turbulent dispersion. The SIMPLE algorithm was used to implement the pressure-velocity coupling with the second-order upwind discretization scheme. The oil droplets were assumed to have a Rosin-Rammler size distribution with a minimum size of 4μm and a maximum of 150, with an average of 40μm. The spread parameter was 1.2 and 10 discrete particle sizes were considered. The heavy oil feed rate is 0.0278kg/s and the oxygen feed rate corresponds to an oxygen ratio of 0.39. The operating pressure used in this model is 1.9MPa. Conductive, convective and radiative heat transfer was accounted for between the gases, droplets and reactor walls. The scattering property of the droplets due to radiation is assumed to be isotropic and the standard P1 model has been used for the purpose of modeling radiative heat transfer. This model is based on the expansion of radiation intensity into an orthogonal series of spherical harmonics. The weighted-sum-of-gray-gases-model (wsggm) was used to calculate the absorption coefficient of the gas phase.

Results and Discussion

Using the chemical kinetics scheme with the parameters outlined in Table 2 - comprising heterogeneous reaction rates from Watanabe et al (2002) and the gas-phase reaction rates of Jones & Lindstedt (1988) and Maki & Miura (1997), we can obtain the following contour plot for the temperature and velocity vector plot within the gasifier (Figure 2 and Figure 3).

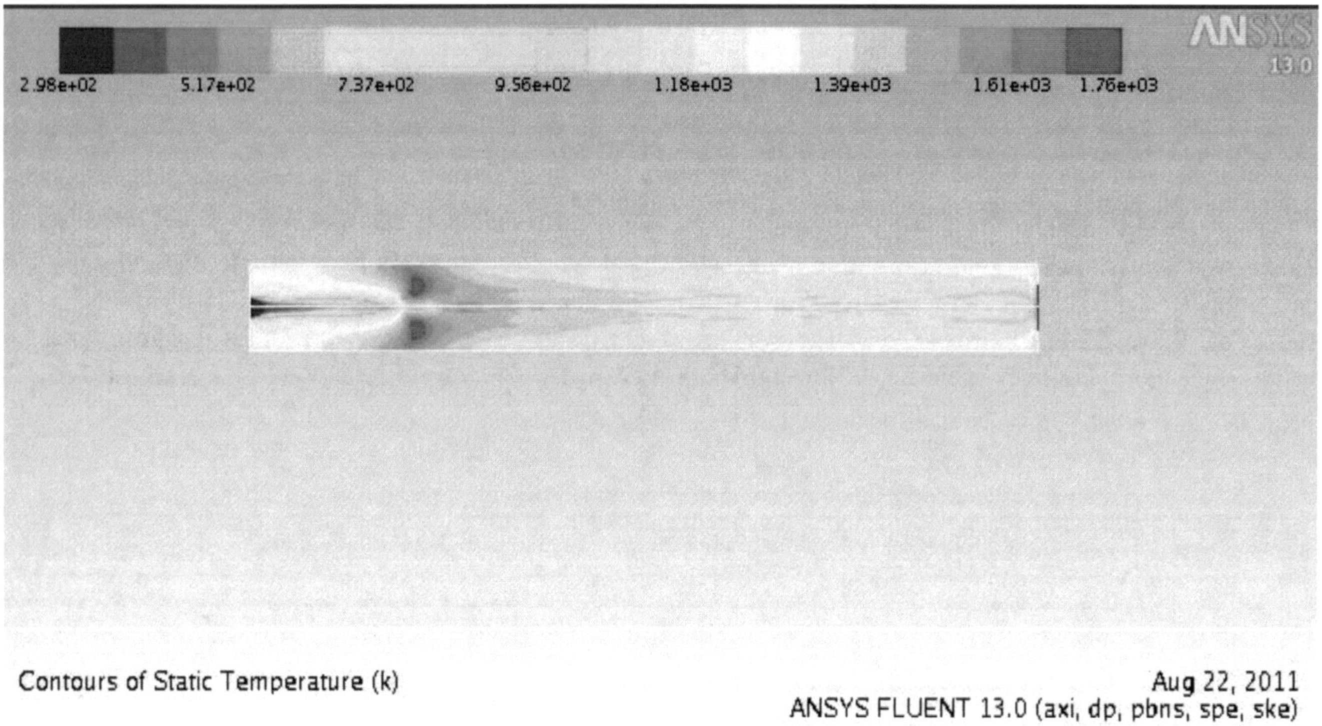

Figure 2: Temperature contour plot

As seen from Figure 2, it can be observed that the maximum temperature within the gasifier is approximately 1760K, and this peak temperature is observed at an axial distance of about 0.6m from the gasifier inlet, which corresponds to the location of the combustion zone within the gasifier. The temperature of the hot combustion gases at the outlet is almost 1330K.

From the velocity vector plot in Figure 3, it is seen that a peak velocity magnitude of approximately 34.4m/s is observed along the flow axis, close to the point of injection of the heavy oil spray.

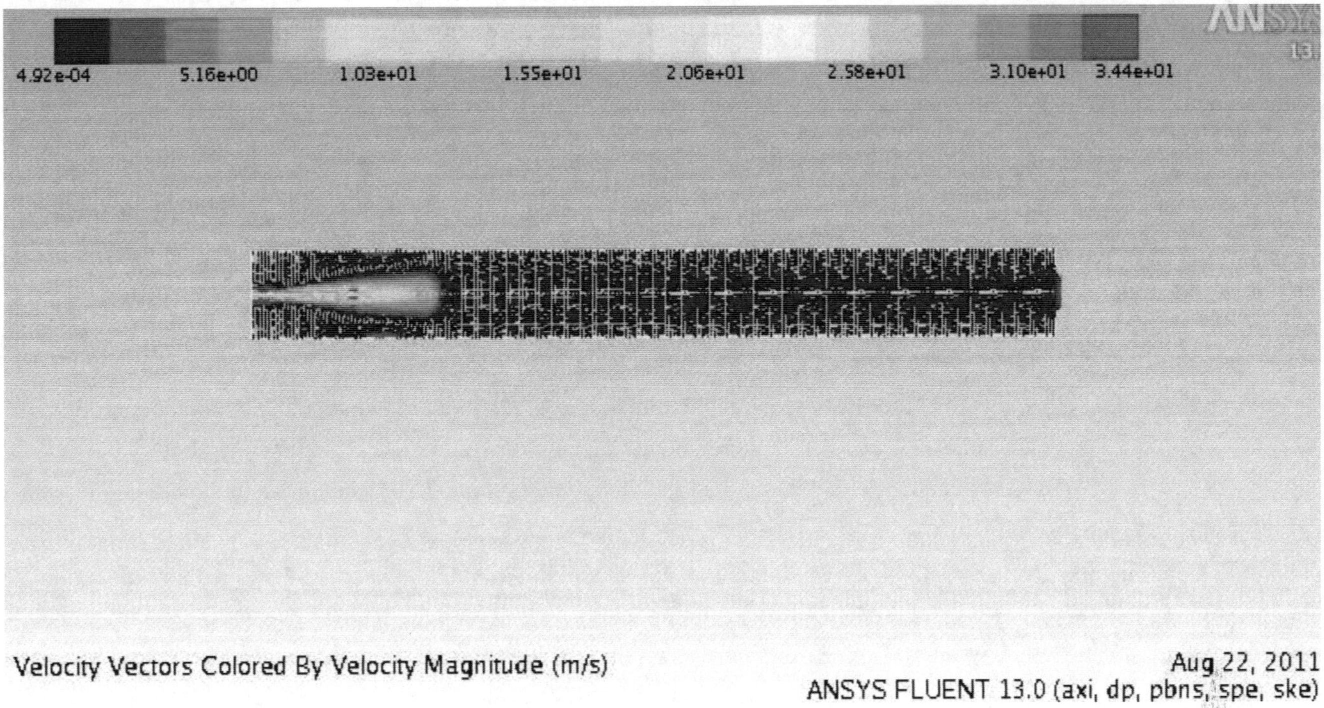

Figure 3: Velocity magnitude vector plot

In Figure 4, the axial temperature profile prediction obtained with the CFD model is compared with experimental data from Ashizawa et al (2005).

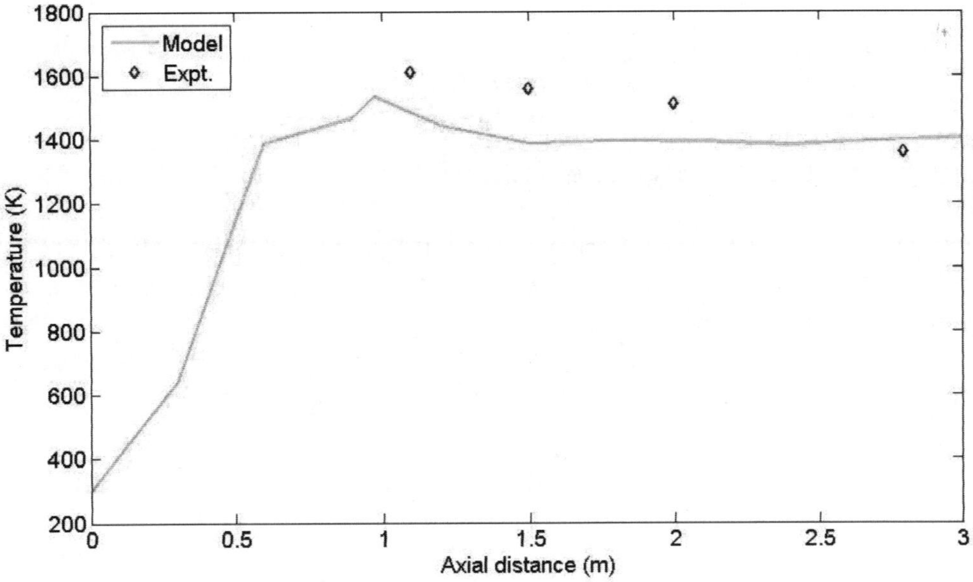

Figure 4: Axial temperature distribution

From Figure 4, it is seen that there is very good agreement between the CFD predictions and experimental results for the axial temperature distribution. The location of the combustion zone as well as the peak temperature is captured well by the model. Furthermore, the temperature trend and magnitude along the axis also shows good agreement with experimental data.

The carbon monoxide and hydrogen species mole fraction model predictions are compared with the experimental results obtained by Ashizawa et al (2005) in Figure 5. Again, it is seen that there is very good agreement with experimental data for both hydrogen and carbon monoxide mole fraction percentages. The correct axial trends are obtained for both the species with the CFD predictions slightly under-predicting the species mole fraction values as compared to the experimental data.

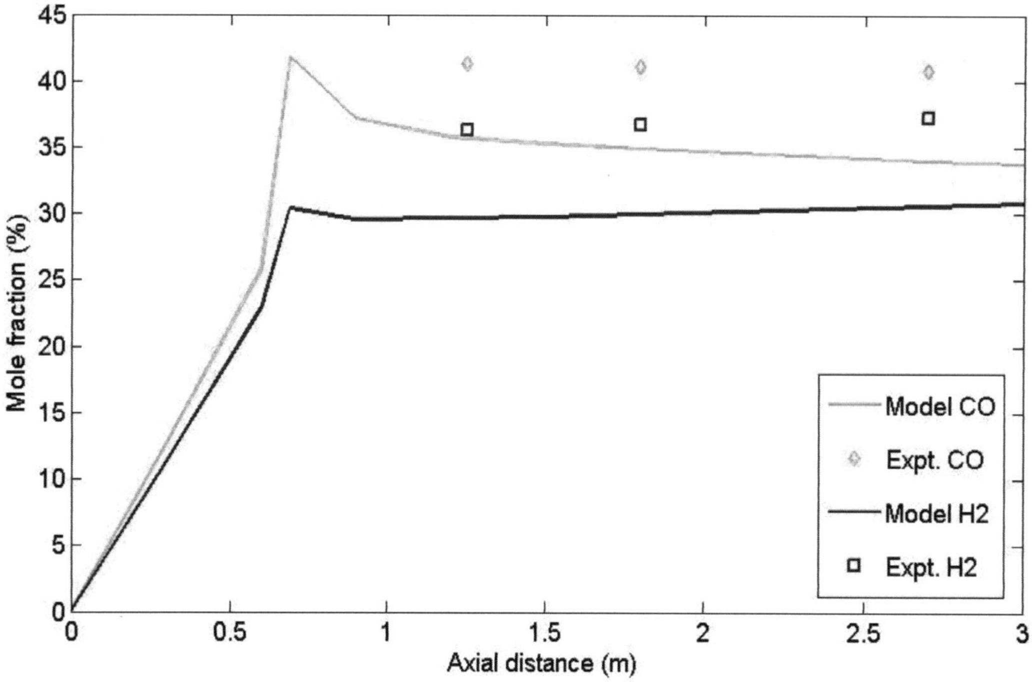

Figure 5: Species mole fraction (%) for carbon monoxide and hydrogen

The carbon dioxide and water vapour species mole fraction model predictions are compared with the experimental results obtained by Ashizawa et al (2005) in Figure 6. It can be observed that the agreement with experimental data for both water vapour and carbon dioxide mole fraction percentages is good. The correct axial trends are obtained for both the species with the CFD predictions slightly over-predicting the species mole fraction values as compared to the experimental data.

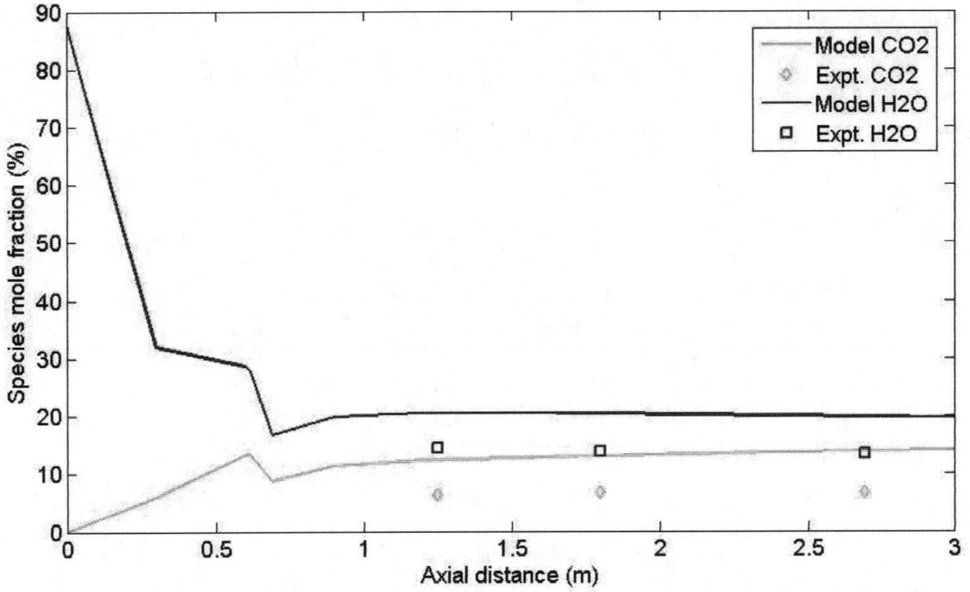

Figure 6: Species mole fraction (%) for carbon dioxide and water vapour

Conclusion

A CFD model for the gasification of heavy oil has been presented in this study which captures the essence of the physical and chemical processes taking place in the gasifier. The accurate chemical composition of the heavy oil (Orimulsion) was obtained from its proximate and ultimate analysis, and the volatiles oxidation chemistry is more detailed than is available in the literature.

From the predictions observed earlier, it is clear that the model captures very well the axial values and trends for temperature and mole fractions of the major species obtained during the gasification of heavy oil. The model slightly under-predicts the absolute values for the axial species mole fractions of carbon monoxide and hydrogen, while slightly over-predicting the values for carbon dioxide and water vapour. Further refinement of the model, including (but not limited to) more detailed combustion modeling for the heavy oil, use of more sophisticated turbulence and radiation models, use of the eddy-dissipation concept, etc. might lead to even more accurate results.

Acknowledgement

The financial support of Statoil Research Centre in Trondheim, Norway is gratefully acknowledged.

Bibliography

ASHIZAWA, M., HARA, S., KIDOGUCHI, K. and INUMARU, J., (2005), "Gasification characteristics of extra-heavy oil in a research-scale gasifier", *Energy*, **30,** 2194-2205.

CHEN, C., HORIO, M. and KOJIMA, T., (2000), "Numerical simulation of entrained flow coal gasifiers. Part I: modeling of coal gasification in an entrained flow gasifier", *Chem. Eng. Sci.*, **55**, 3861-3874.

HIGMAN, C. and VAN DER BURGT, M.,"Gasification", Second Ed. 2008, Elsevier Science.

JONES, W.P. and LINDSTEDT, R.P. (1988), "Global Reaction Schemes for Hydrocarbon Combustion", *Combust. Flame*, **73**, 233-249.

MAKI, T. and MIURA, K. (1997), "A Simulation Model for the Pyrolysis of Orimulsion", *Energy & Fuels*, **11**, 819-824.

SHI, S.-P., ZITNEY, S.E., SHAHNAM, M., SYAMLAL, M. and ROGERS, W.A. (2006), "Modelling coal gasification with CFD and discrete phase method", *J. Energy Inst.*, **79(4)**, 217-221.

WATANABE, H., OTAKA, M., HARA, S., ASHIZAWA, M., KIDOGUCHI, K. and INUMARU, J. (2002), "Modelling and Simulation for Extra Heavy Oil Gasification on Entrained Flow Gasifier", *Proc. IJPGC'02*, 667-674.

SMR3D: AN INDUSTRIAL CFD TOOL FOR LARGE STEAM METHANE REFORMER DESIGN OPTIMIZATION

Dr. Julien Cancès, Air Liquide R&D, Paris, France

Frédéric Camy-Peyret, Air Liquide R&D, Paris, France

Dr. Dieter Ulber, Lurgi GmbH, Frankfurt am Main, Germany

Introduction

The worldwide hydrogen demand has strongly increased in the past years, mostly for refinery needs in order to match environmental requirements such as desulfurization. This trend is expected to maintain for a long period and raise new challenges associated to the design and operation of large scale hydrogen plants. As of today most hydrogen plants are based on steam reforming of natural gas [1]. Steam reforming is also applied to produce CO or synthesis gas for other products.

As a major actor in the field of hydrogen production, and hydrogen plant engineering through its subsidiary LURGI, AIR LIQUIDE has developed advanced in-house simulation tools to understand, predict and optimize the efficiency, compactness and reliability of large scale steam methane reformers.

The present paper aims at presenting the SMR3D modeling tool and its applications to key aspects of a recent LURGI reformer design.

SMR3D presentation

SMR3D consists of a 3D CFD tool, ATHENA, dedicated to large industrial combustion chambers simulations coupled with a 1D SMR tube model.

Combustion chamber model

ATHENA, the AIR LIQUIDE in-house CFD code developed since the 80's, is a robust industrial tool, which includes several established models dedicated to heat transfers in industrial furnaces, including oxy-combustion [2]:

- The averaged three-dimensional flow is solved by use of a RANS approach (Reynolds Average Navier-Stokes). The Navier-Stokes equations are computed averaged on time, while turbulent fluctuations are modeled with using the standard $k\text{–}\varepsilon$ model [3].
- The solver is based on a cell-centered, structured finite volume discretization of Navier-Stokes equations [4] and on transport equations of passive or active scalars (enthalpy, mass fractions, turbulence quantities of RANS models ...), with a SIMPLE-algorithm segregated solution procedure.
- Radiative heat transfer equations are solved by use of a specific ray tracing method (DTRM: Discrete Tracing Radiative Method [5]), a method which proved to be adapted to complex geometries. The gas radiation properties are taken into account with a Weighted Sum of Grey Gases (WSGG) model with four grey gases tables available for air and oxy-combustion products of different fuel compositions [6], with or without soot. The high radiation contribution to SMR furnaces heat transfer implies the use of an efficient and detailed radiation model, taking into account the geometric complexity of industrial reformers.
- The Magnussen eddy dissipation turbulent combustion model [7] is commonly used for industrial furnaces simulations, because it provides several pertinent tuning parameters thus enabling the use of a coarse mesh suited to such large geometries and avoiding to fully detail the burner geometry.
- The heat losses at the combustion chamber walls are evaluated using a global heat transfer law, as described in the following equation:

  $$\varphi = K_{wall} \cdot (T_{wall} - T_{\infty})$$

  The wall heat transfer coefficients – K_{wall} – were calculated by means of the thermal resistance method using the material properties provided by vendors. The heat conduction is computed inside the tunnel walls using an energy coupling at the interface.

Reformer tubes model

The computation of the flow, heat transfers and chemical endothermic reaction of steam methane reforming inside the catalytic tubes has been carried out using the code developed at Gent University described in [8][9]. The tube is described as a 1D plug-flow reactor in the vertical direction and computes the Hougen-Watson reactions set:

$$\begin{aligned} CH_4 + H_2O &\Leftrightarrow 3 \cdot H_2 + CO \\ CO + H_2O &\Leftrightarrow H_2 + CO_2 \\ CH_4 + 2 \cdot H_2O &\Leftrightarrow 4 \cdot H_2 + CO_2 \end{aligned}$$

The reforming model takes into account the intrinsic catalytic kinetics and the limitation through diffusion into the porous catalysts, using reaction effectiveness factors. The effectiveness factors can be computed by the gent code, however these parameters were considered as constant in order to speed-up the simulations and their values were established on the basis of operational data gathered on AIR LIQUIDE operated plants.

The coupling between the combustion chamber and tubular reactor model is performed by iterative computations in the firebox followed by reforming reactions in the tube. The heat flux computed in the combustion part is imposed as a boundary condition of the tube model, which returns the tube temperature by classical energy balance.

Numerical implementation

The SMR3D solver has been developed in order to take advantage of large computers. Two levels of parallelization (MPI and OpenMP) allow running massively parallel computations involving hundreds of million cells with a reasonable restitution time. Therefore, simulation of a full scale large reformer with hundreds of tubes and burners is now at hand. An example of typical SMR3D computation results for a large furnace geometry is illustrated in Figure 1 (Note that due to symmetry, only half of the furnace is computed).

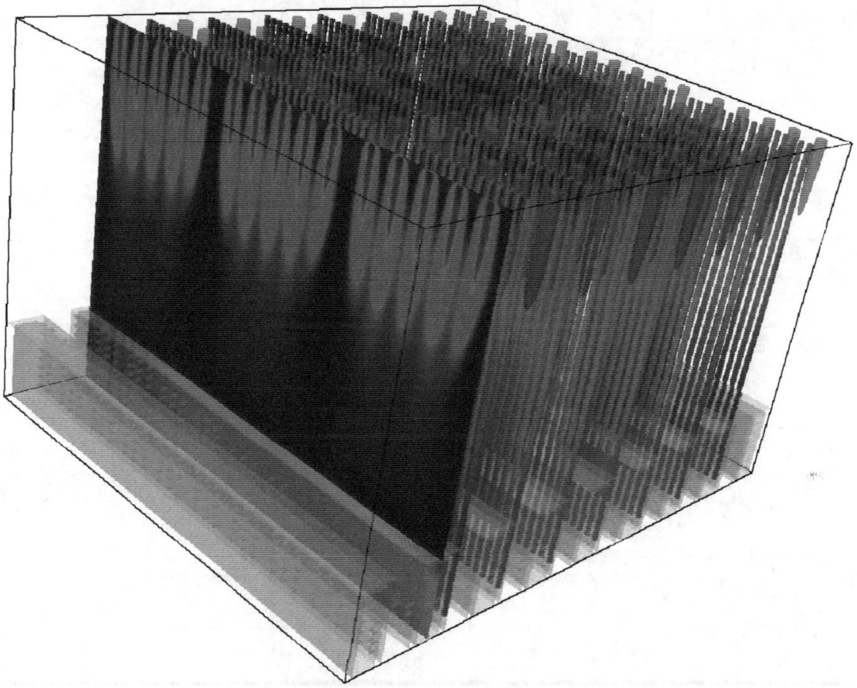

Figure 1: SMR3D computation results illustration for a large SMR furnace. The tubes are colored by temperature. Flames are represented as iso-velocity surfaces at 20 m/s.

SMR3D validation

Before using SMR3D as a design tool, it was necessary to ensure the capabilities of SMR3D to predict accurately the behavior of industrial size steam methane reformers. Therefore, SMR3D has been validated through confrontation with experimental data measured on LURGI design fireboxes worldwide. As an illustration the validation protocol applied to a recent CO plant based on LURGI technology is presented in the present paper.

Mass and Energy balances

The first model validation step is to compare the overall heat and mass balances with the ones gathered on-site using available measurements. However, as each sensor is assorted with a certain uncertainty it is necessary to close balances by applying a post-treatment step to the raw data, called data reconciliation [10], in order to get robust and reliable overall mass and energy balances. The data reconciliation method is a statistical approach combining measurement tolerances with a process simulation, which provides the most probable value - associated with a range confidence plotted as error bars in following diagrams. Thermal losses have been characterized as well during the field test campaign through external temperature measurements which have been analyzed using ASTM 680 standard [11].

Table 1. Comparison of the heat and mass balance for a CO plant experimental (reconciled) data and SMR3D computation results.

	Experimental data		**SMR3D**	
	Mass [kg/s]	Heat [MW]	Mass [kg/s]	Heat [MW]
Injected at burners	75.34	**216. 51** (100.0%)	75.34	**215.01** (100.0%)
Transferred to tubes		**115.12** (53.2%)		**115.14** (53.6%)
Radiation portion				111.92 (97.2%)
Reformed gas temperature		*914°C*		*917°C*
Flue gas	75.34	**99.26** 45.8%	75.34	**98.91** 46.0%
Temperature		*1036°C*		*1020°C*
Heat losses		**2.14 (1.0%)**		
Tube feeders		0.84[1] (0.4%)		
Firebox walls		1.28[1] (0.6%)		**0.95** (0.4%)

[1]: determined using ASTM 680 standard

The overall mass and energy balances are gathered in Table 1 and illustrated in Figure 2, whereas the flue gas and syngas temperature are plotted in Figure 3. A very close agreement between experimental and simulation results is observed. The SMR3D simulation results are inside the confidence intervals. The experimental heat losses evaluation encompasses also the tube feeders' contributions. As the SMR3D value (0.95 MW) corresponds to the furnace walls heat fluxes, it has to be compared to the firebox heat loss estimation (1.28 MW) giving a closer match.

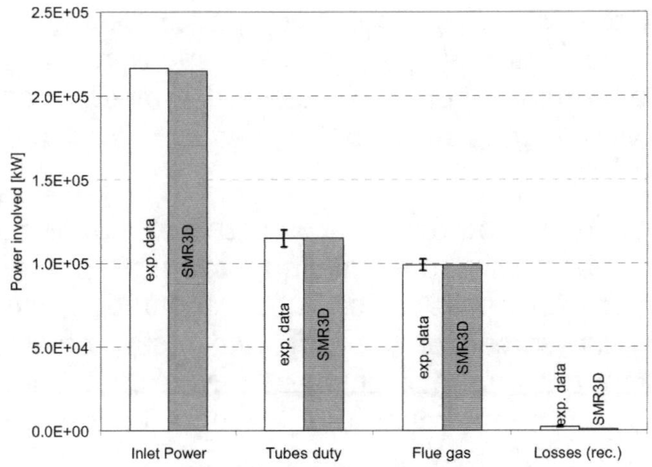

Figure 2. Comparison of the main terms of the energy balance for experimental (reconciled) data and SMR3D computation results.

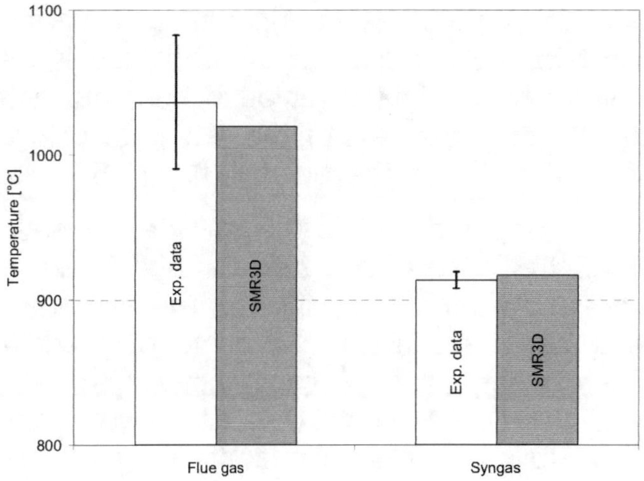

Figure 3. Comparison of the syngas and flue gas outlet temperatures for reconciled data and SMR3D computation results.

A comparison between experimental – i.e. reconciled – and computed syngas composition at the tubes outlet is shown in Figure 4. Both results are very close and comparable one to another, always within the error bars.

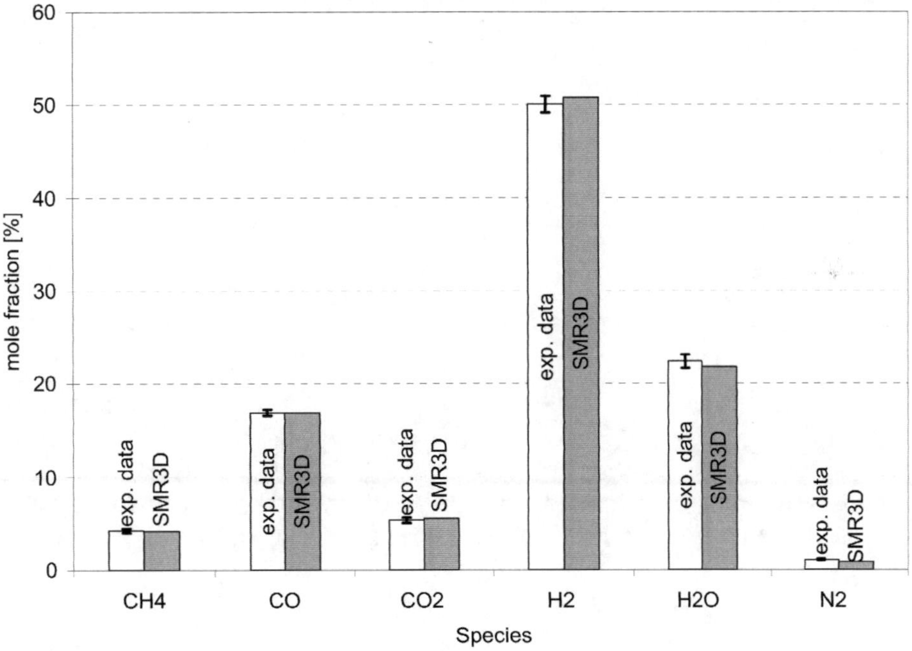

Figure 4. Comparison of the syngas outlet composition for reconciled data and SMR3D computation results.

Tubes temperature profiles

The second step of the validation approach is to compare the energy and flow distribution inside the reformer with experimental data. In the present case of tubular reformer, the tube skin temperatures were measured with a pyrometer applying wavelengths such that the influence of gas radiation is minimized. As this type of device measures and converts the incoming radiative heat flux into a temperature, the measured temperatures require post-treatment of the raw data using a view factors method in order to correct the reflected part from the radiative flux [15].

Computed and corrected experimental tube temperatures are compared in Figure 5. The SMR3D average over all tubes and the tubes with the minimum and maximum temperature profile are shown. The average temperature profiles for both computation and experiments are in close agreement. In addition, the spread of the corrected measurement values is only in a few cases not within the computed min/max envelopes. Note that both SMR3D and experimental data agree well with the maximum temperature taken for the design of the tubes and calculated at design stage with a commercial tool adapted to LURGI's reformer design.

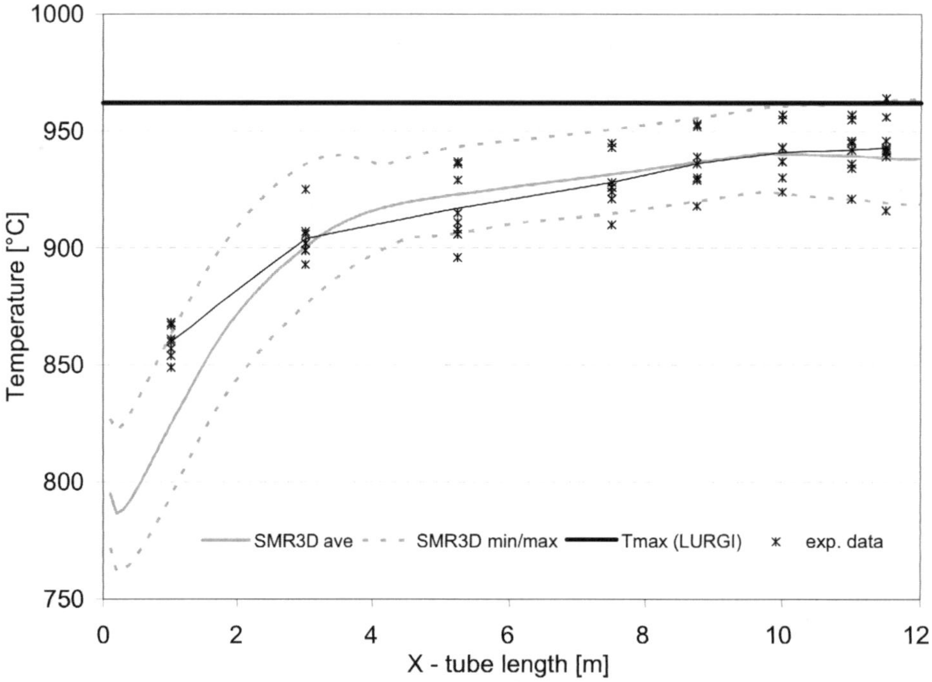

Figure 5. Tube skin temperature profile: corrected experimental data (symbols), SMR3D computation results (average – solid line; min/max envelop – dotted line) and LURGI maximum temperature calculation.

SMR3D applications

Based on the SMR3D predictions and its confrontation with experimental data for overall furnace mass and energy balances and for tube skin temperature profiles as layed down in the preceding paragraph the prediction capability of SMR3D not only allows to describe and understand the governing physical phenomena inside the firebox but also to exploit the detailed coupling effects between the combustion chamber and the reforming tube, with a level of details not reported, to our knowledge, in the literature so far.

A few examples of the application of this tool for design optimization and further improvement of the understanding of the technology are given below.

Heat losses and refractory materials

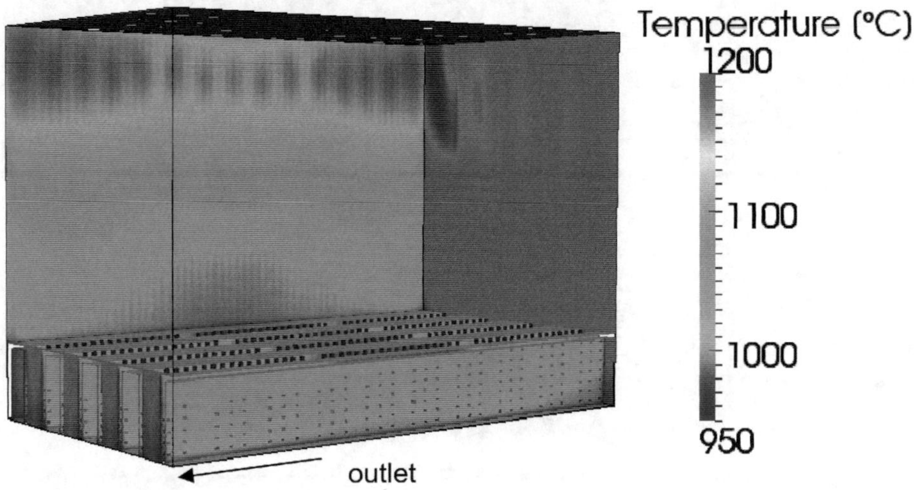

Figure 6. Furnace walls temperature distribution illustration for a large SMR furnace.

A heat flux or temperature mapping at the firebox walls, as illustrated in Figure 6, enables to characterize and map precisely the heat losses of the furnace. The upper part of the reformer walls, subjected to the flames, is the main contributor to heat losses. Additionally, a precise and reliable prediction of the walls temperatures, depending on the operating conditions, helps in the refractory and insulation materials definition during the design phase.

Flow arrangement

The flow arrangement inside the combustion chamber could be characterized and precisely analyzed. In the velocity magnitude field, pictured in **7**, bending of the flame from the external burner rows is observed to a much lower extent than that described in [12].

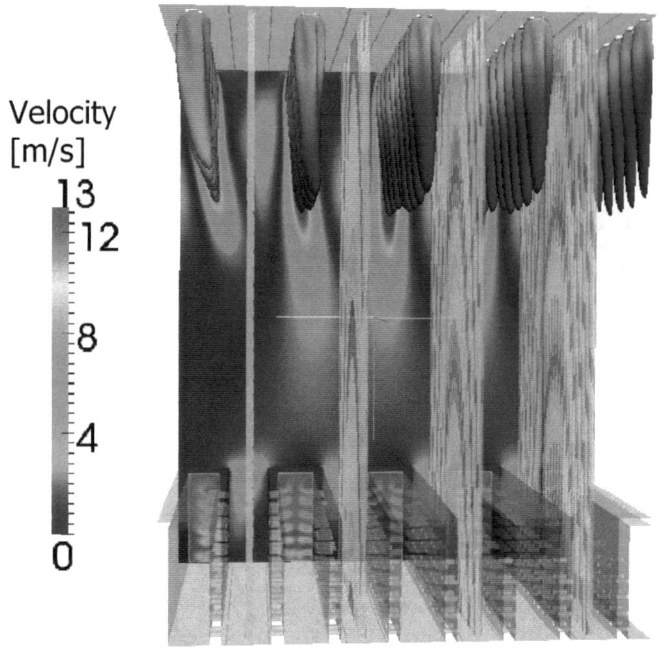

Figure 7. Velocity magnitude field illustration. Flame are represented by iso-temperature surfaces at 1500°C coloured with velocity.

Tube duty polar distribution

The normalized heat flux, $\tilde{\varphi}(\theta,z)$ and temperature variation, $\Delta T(\theta,z)$, circumferential profiles at different altitudes are plotted in Figure 8 for a given 5″ tube with 300 mm pitch. Definitions for heat flux normalization and temperature variations used here are given below. The tube is heated on both sides by radiation emitted from hot flue gas and flames, whereas it receives no energy from the neighboring tubes that have locally almost the same temperature. Consequently, the normalized heat flux and temperature variations profiles show two minima in the angular sectors that face other tubes. The ratio between the maximum value for $\tilde{\varphi}(\theta,z)$ of 1.11 computed using SMR3D is close to the API 530 view factor prediction of 1.18, which is a value that doesn't take into account the convection and conduction diffusion influence [13].

The maximum temperature deviation from the average value at the top of the tubes is around 15°C. Indeed, the heat flux received is high in front of the flames, however the syngas and tube wall temperature is low at this location. The temperature difference at the bottom tube section where the maximum tube wall temperature occurs is even lower because of the lower heat flux, an inherent feature of a top-fired reformer enhancing tube lifetime.

$$\tilde{\varphi}(\theta,z)=\frac{\varphi(\theta,z)}{\frac{1}{2\pi}\int_{2\pi}\varphi(\theta,z)d\theta} \qquad \Delta T(\theta,z)=T(\theta,z)-\frac{1}{2\pi}\int_{2\pi}T(\theta,z)d\theta$$

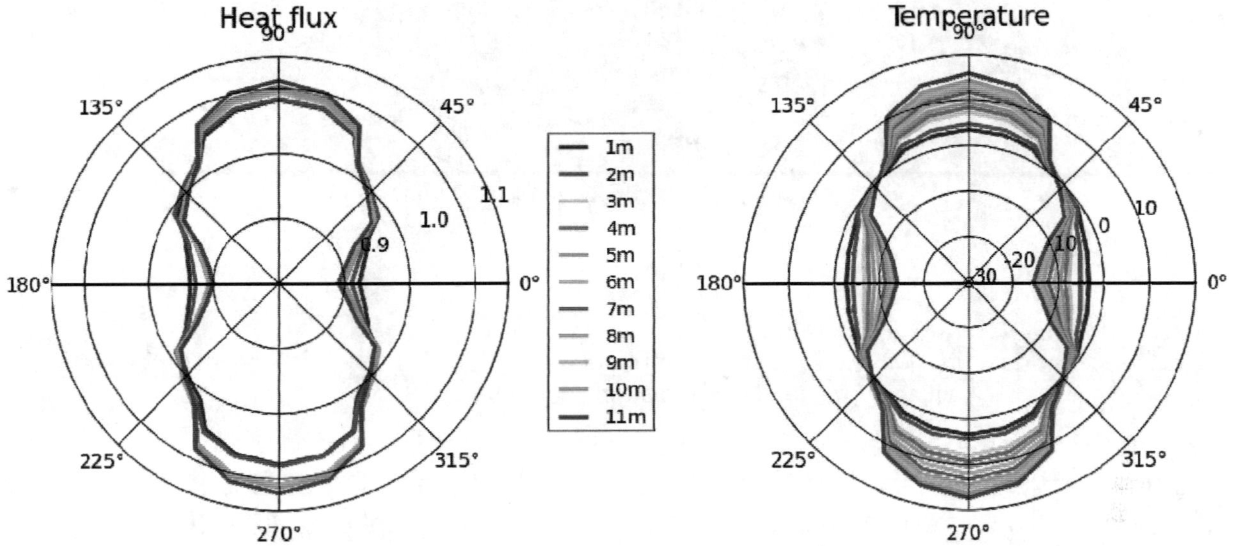

Figure 8. Distributions of the normalized heat flux and temperature difference to the average for a given 5″ tube at different altitudes (from the bottom).

Tunnel ports effects on tubes

A potential negative tunnel effect on tube lifetime has been proposed by Farnel on the basis of Monte-Carlo simulations: the radiation intensity could be intensified by multiple reflection inside the tunnel and impact the tubes placed right in front of the opening, inducing a local heat flux increase by 20%, yielding to a local temperature increase of 15°C [14]. The detailed radiation heat transfer model implemented within SMR3D, including finely discretized ray tracing and precise gas radiating properties, has been used in order to characterize this phenomenon. The tunnels area as been highlighted in Figure 9 for temperature fields and radiation contributions (two color codes: tube surface heat flux (W/m^2) and gas radiation source term (W/m^3)). Neither radiation intensification nor temperature peaks have been observed locally on the tubes. On the contrary, the tunnels protect the tubes, as tubes temperature decreases in the tunnel region (see Figure 5). Moreover, it reveals that the tunnels have a small cooling effect on the flue gas entering the tunnels, as illustrated by the temperature gradient between both sides of the tunnel wall in Figure 9. The heat is transferred to the walls and then conducted to the bottom. In terms of radiation, the inside of the tunnel tend to absorb (red) the radiation coming from the outside (green).

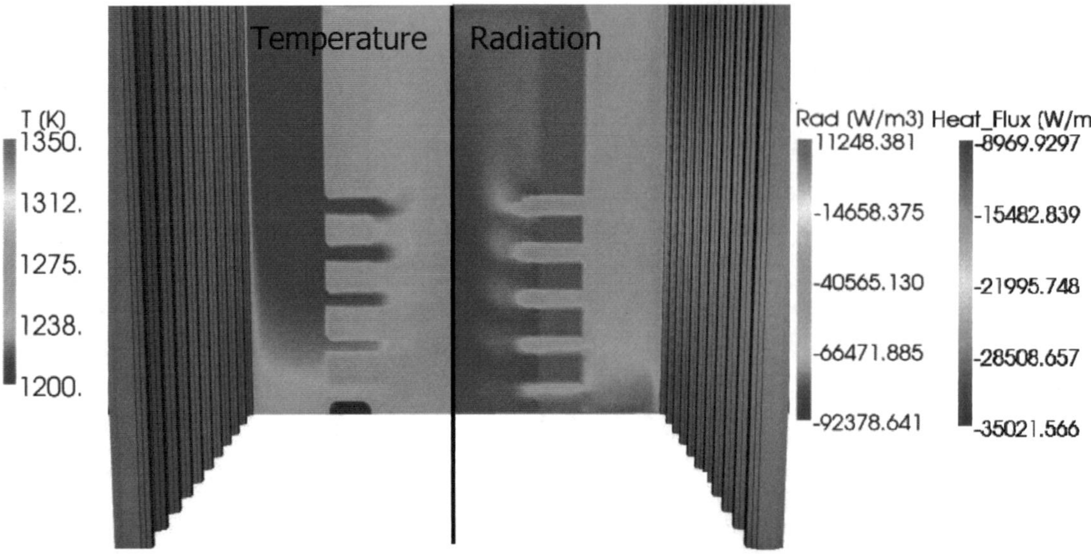

Figure 9. Temperature (left) and radiative contributions fields (right) (two color codes: tube surface heat flux (W/m^2) and gas radiation source term (W/m^3)).

As a result from the detailed coupled computations presented here, the radiation intensification such as in a laser device observed in [14] appears unrealistic in a SMR furnace tunnel. However the cited paper documents several assumptions like the Monte-Carlo approach is not rigorous as detailed in classical references [15][16], the 3D problem is calculated in 2D, spectral aspects from gas, tubes and bricks are neglected, tunnel and tubes walls with specular reflection and uniform gas temperatures assumption is unrealistic and the gas emissivity is twice as big as values given in other literature, e.g. [17].

Design improvement

Finally, the continuous improvement of the commercial LURGI reformer is benefitting from SMR3D simulation capabilities. As an illustration, two of the largest AIR LIQUIDE operated plants, recently designed by LURGI, were simulated and the results are compared (~400 tubes on 8 rows each). The so-called HERACLES plant has been commissioned in 2011 (Figure 10(a)) and was designed 2 years ago, whereas the second project will be started-up in 2013 (Figure 10(b)). Therefore, the latter can benefit from SMR3D inputs to optimize the heat flux and temperature homogeneity and the flue gas flow arrangements. So SMR3D allows significant performance improvements described further below.

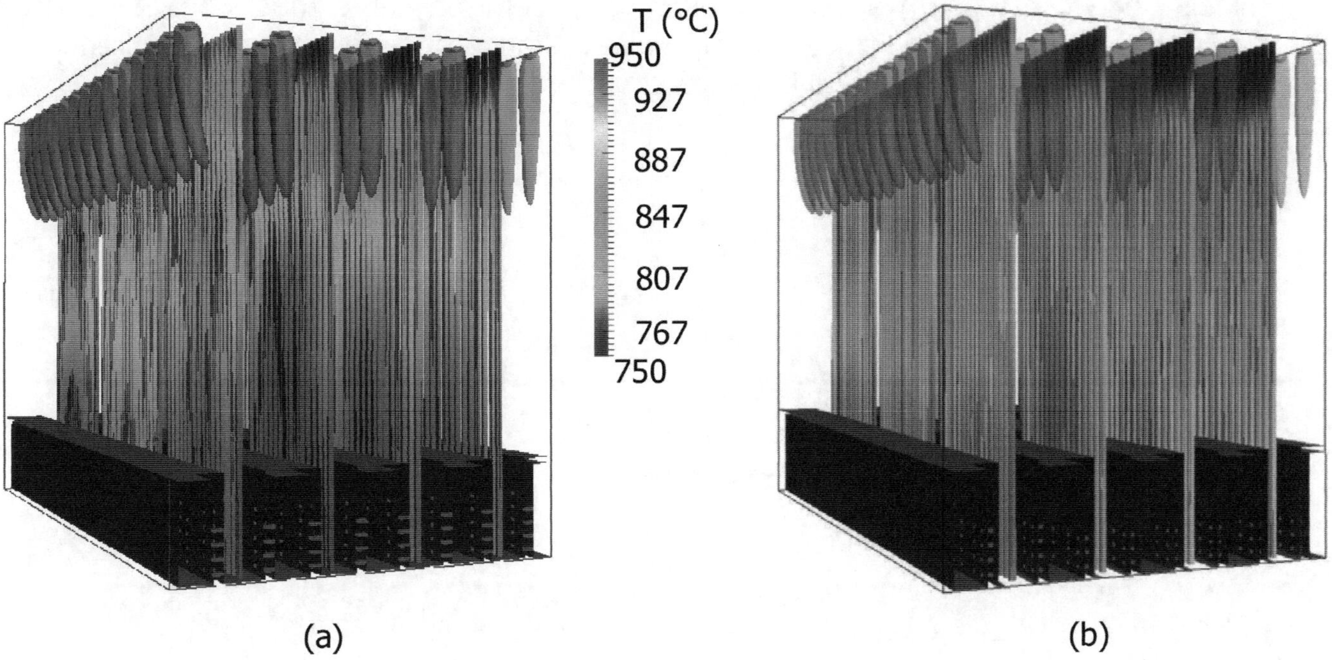

Figure 10. Illustration of the computed results for two of the largest AIR LIQUIDE operated furnaces: (a) HERACLES, (b) undisclosed project

As the two plants do not have exactly the same design characteristics and operational conditions (e.g. 4″ vs. 5″ tubes, different feedstock, etc.), normalized physical units are compared. The normalized tube duty is plotted in Figure 11 for each tube, row per row. HERACLES data is illustrated using void symbols, whereas the other plant is represented by grey bold symbols. Reforming efficiency rises with higher temperature, while the maximum operating tube temperature is limited by the design for economically optimized reforming conditions and a certain tube lifetime based on creep rupture strength curves. As the hottest tube will limit furnace operation, homogenization of the tube duty at the furnace scale is key in improving furnaces design and operation.

For firebox structural reasons, the tube rows in both furnaces have been split into 3 tubes section. The SMR3D computation results allowed to observe a tube duty maldistribution for HERACLES, which had to be compensated by an adjustment of the firing. Less heat would have been transferred to the central section than that to the extremity ones. This effect was due to different levels of power available per tubes locally, and has been corrected in the latest design.

The average trends have been added to the scatter plot in Figure 11, showing a slight increase of the average duty with an increase of tube number for the HERACLES case. A detailed CFD investigation was performed to identify the reasons for such tendency. It has been found that the flue gas flow arrangement in the firebox was responsible for this inhomogenity. This was corrected in the latest reformer design thanks to an optimized tunnel openings distribution, allowing a homogeneous aspiration of the flue gas in the combustion chamber, while minimizing the pressure induced.

Finally, between these two generations of large reformers, the tube loads have also been significantly homogenized, since the normalized standard deviation has been reduced from 2.3% to 1.2% from HERACLES to the second project.

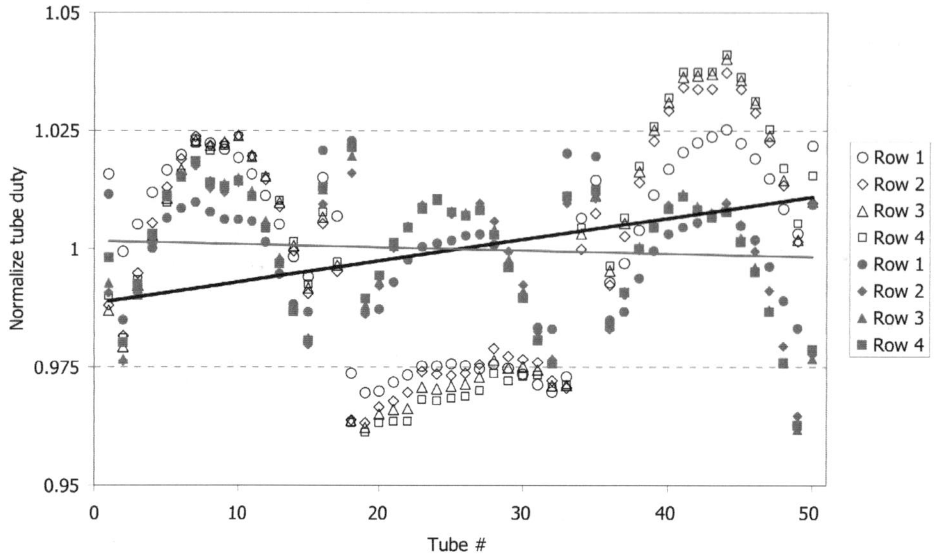

Figure 11. Comparison of normalized tube duty distribution for HERACLES (void symbols) and second project (grey symbols). Row 1 corresponds to the side wall, row 4 is in the furnace middle.

Conclusion

SMR3D is the result of 25 years of continuous efforts for rigorous integration of the most appropriate models, while keeping the solver highly efficient. It is now possible to run simulations of full scale large SMR furnace, including combustion, radiation and the coupling of combustion chamber to individually calculated reforming tubes.

The SMR3D capability to describe different types of reformers (H_2 or CO production plants, 4 or 5 inches tubes) demonstrates its flexibility. The return of experience from SMR3D computations has already provided successful improvement of industrial SMR furnaces.

The understanding of key phenomena involved in reformers operation and design has been greatly improved, resulting in further optimization and benefits in terms of efficiency, reliability and cost optimization.

SMR3D paths the way for larger steam reformer designs, where homogeneity and equal-distribution is essential for the operation of highly reliable and available plants.

References

[1] New York Power Authority; (2005), "Hydrogen fact sheet: Hydrogen production – Steam Methane Reforming (SMR)", www.getenergysmart.org

[2] Champinot, C., Till, M., Haung-Naudin, C., Klug, A., (1996), "Gekoppelte Berechnungen von Glas und Atmosphäre", *Glas Ingenieur*

[3] Jones, W.P., Launder, B.E., (1972), "The prediction of laminarization with a two equation model of turbulence", *Heat and mass transfer*, 15, pp. 310-314

[4] Patankar, S.V. (1980), "Numerical Heat Transfer and Fluid Flow", *Hemisphere*.

[5] Lockwood, F.C. and Shah, N.G., (1980), "A new radiation solution method for incorporation in general combustion prediction procedures", *18th Symposium (International) on Combustion, The combustion Institute, Pittsburg,* pp. 1405-1414

[6] Soufiani, A., Djavdan, E., (1994), "A comparison between weighted sum of grey gases and statistical narrow band radiation models for combustion applications", Combustion and Flame, Vol. 97, pp. 240-250

[7] Magnussen, B.F., Hjertager, B.H., (1977), "On Mathematical Modeling of Turbulent Combustion with Special Emphasis on Soot Formation and Combustion", *16th Symp. (Int.) on Combustion,* pp. 719-729

[8] Xu, J., Froment, G.F., (1989), "Methane Steam Reforming – I Methanation and Water-Gas Shift: Intrinsic kinetics"; *AIChE journal*, 35(1), pp. 88-96

[9] Xu, J., Froment, G.F., (1989), "Methane Steam Reforming – II Diffusional limitations and reactors simulation"; *AIChE journal*, 35(1), pp. 97-103

[10] Kalitventzeff, B., Cransfeld, J.M., Vassilakis, C.E., Dimaratos, J., Delava, P., (1999), "Demonstration of an Energy Saving Process for a Crude Oil Distillation Unit", ESCAPE 9, Budapest, (Hungary), June.

[11] ASTM C680-08, (2008), "Standard Practice for Estimate of the Heat Gain or Loss and the Surface Temperatures of Insulated Flat, Cylindrical, and Spherical Systems by Use of Computer Programs", *ASTM International*

[12] Cotton, W., (2003), "Fluegas flow patterns in top-fired steam reforming furnaces", *Johnson Matthey Catalysts, IMTOF.*

[13] API 530 standard, 2008, (2007), "Calculation of Heater-tube Thickness in Petroleum Refineries", identical to ISO 13704:2007

[14] Farnell, W., (1999), "The Tunnel Port Effect: Validation by Monte Carlo Simulation", *Synetix ICI Group, IMTOF.*

[15] Siegel, R., Howell, J.R., (1981), "Thermal Radiation Heat Transfer – Second Edition", *HPC New York*

[16] Modest, M.F., (2003), "Radiative heat transfer – Second Edition", *Academic Press*

[17] Truelove, J.S., (1976), "A mixed grey gas model for flame radiation", *AERE Harwell*

(389a) A Suitable Disengagement: Adventures In a Multi-Phase Reaction System

Cheryl Teich, Core R&D/ Process Chemistry & Development, The Dow Chemical Company, Spring House, PA

ASuitable Disengagement: Adventures in a Multi-phase Reaction System

CherylTeich
The Dow Chemical Company, Spring House, PA
Core R&D/Process Chemistry and Development

In the early business development of a niche specialtyproduct, a small-scale CSTR provided adequate volumes. The product is generated as a gas that forms througha complex reaction pathway and must disengage from a mixture containing twodistinct solids and multi-component liquid and gas phases. Extended productholdup can lead to yield loss via byproduct and mer formation, and even to apotentially hazardous condition. Safe and successful operation of the processrelies on balancing the reaction pathway with adequate reactor mixing and gasdisengagement. This had been achieved with the existing process throughextensive operating experience.

The potential for step growth in the business led to anexamination of reactor options to supply the larger volumes. A concern wasthat a larger CSTR might not provide adequate gas disengagement and that theprocess would suffer as a result. The first step in generating the nextgeneration reactor design was to capture process understanding in the processscheme shown below.

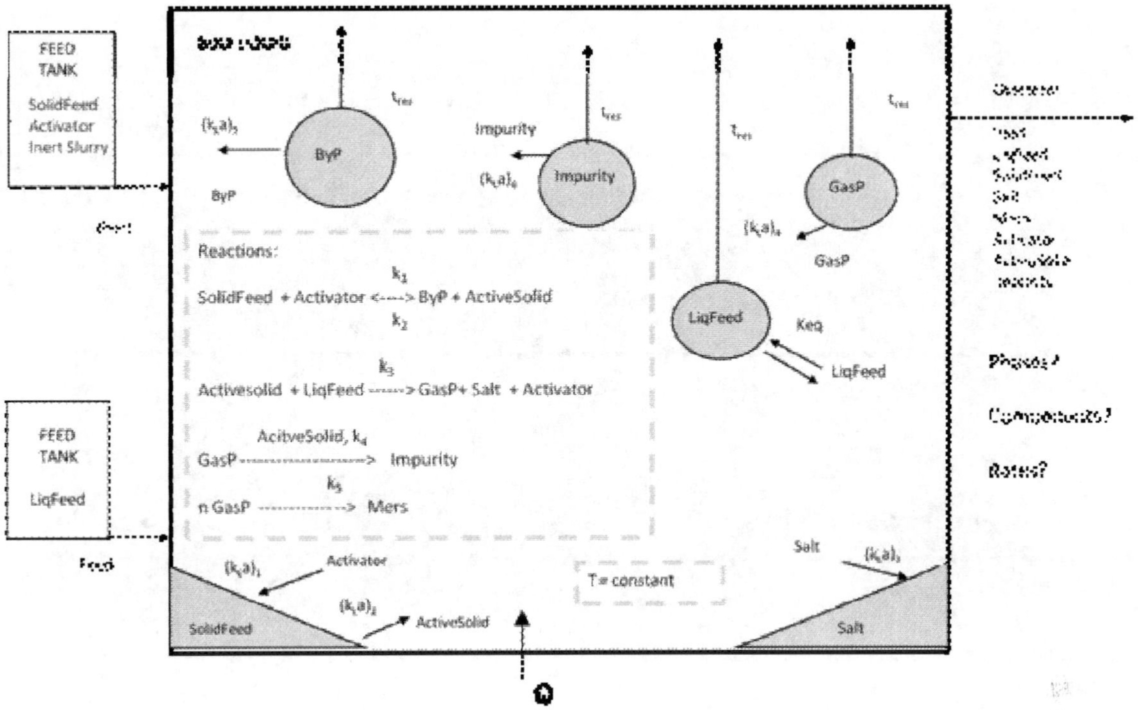

Development of the process scheme led to an examination of thescale-independent chemistry and its interaction with equipment-specificagitation configurations. The development program used various laboratoryreactors and modeling tools such as DynoChem and Aspen Plus to deliver volume-specificrecommendations on reactor type and designs to meet the growing business needs.

See more of this Session: Mixing In the Process Industries - How to Develop a Successful Process
See more of this Group/Topical: North American Mixing Forum

SABIC
Innovative
Plastics™

Sharing our futures

2

Introduction

- Scale-up of an air oxidation reaction to commercial scale was required for a plant expansion by Sabic Innovative Plastics.
- Heat and mass transfer considerations were critical to the success of the scale-up.
- The agitator design was a significant key to success:
  - Sufficient heat transfer to maintain the pilot plant cycle time
  - Mass transfer at least as good as seen at the pilot scale
- Agitation tests were performed in the vendor's facility to trial the proposed design and to optimize the performance.

SABIC
Innovative
Plastics™

Air Oxidation Process - Reaction

$$\text{organic substrate} + O_2 + N_2 \xrightarrow[\text{diluent}]{\text{catalyst}} \text{product} + N_2$$

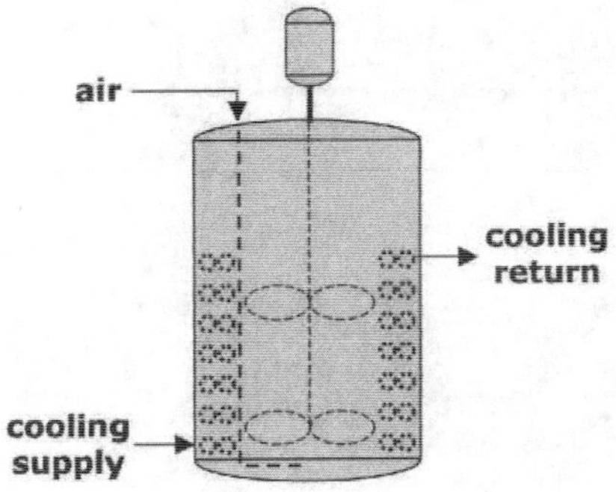

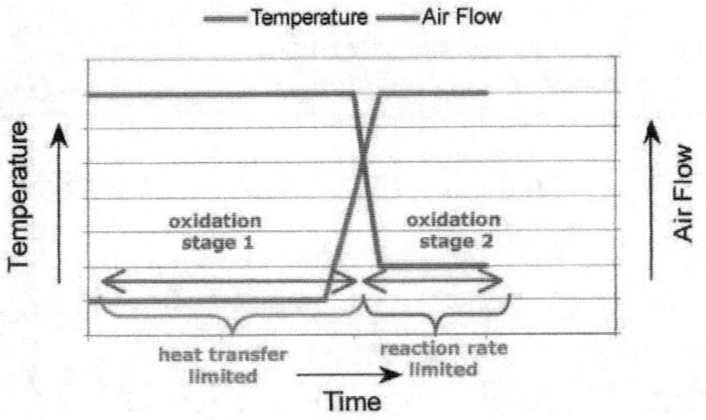

20X Scale-Up Required From Pilot Scale

SABIC Innovative Plastics™

سابك
sabic

Scale-Up Considerations

Critical for Scale-Up

- Heat transfer (cooling)
  - rate limiting during oxidation reaction
  - jacket surface to Rx volume ratio significantly lower @ commercial scale vs. pilot scale
- Mass transfer
  - not rate limiting at the pilot scale
  - need at least equivalent mass transfer capability at the commercial scale
- Optimization of the proposed design
  - baffling required with the presence of cooling coils
  - location of impellers vs. point of air introduction
  - good distribution of air bubbles

SABIC Innovative Plastics™

سابك
sabic

Agitation Hydrodynamic Study

A hydrodynamic study was commissioned in the agitator vendor's facility for:

- optimization of baffling, impeller placement, and air distribution
- mass transfer measurements
- heat transfer measurements
- mixing time analysis
- comparison with previous piloting configuration

- 1 meter ID transparent vessel
- Medium = water @ ambient T, atm. P
- Geometrically scaled from the proposed commercial design
- Internal coils with heating/cooling capability

Design Optimization

Baffle optimization:

- Baffle configuration modified from initially proposed design.
- Vortex eliminated and full theoretical power input demonstrated

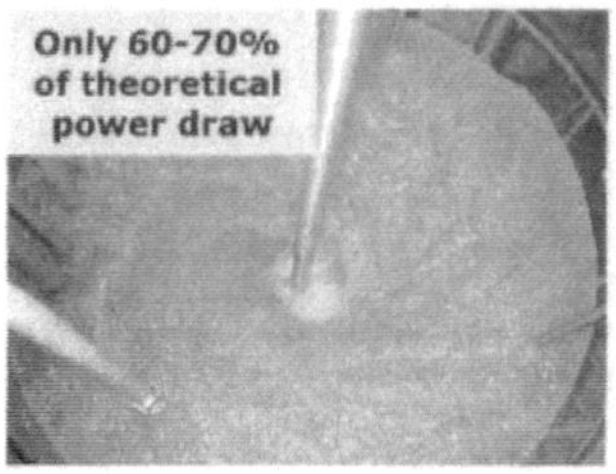

Before baffle changes

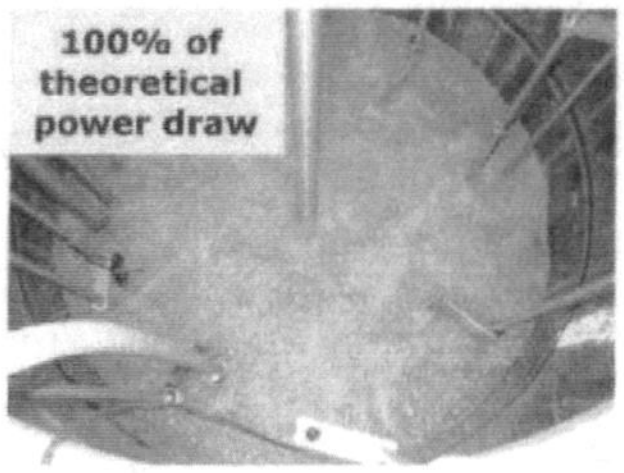

After baffle changes

Lower impeller elevation:

- Originally positioned to maximize heat transfer in the adjacent coils
- Lowered to improve overall performance
- Vertical fluid velocity between the coils increased from 0.18 to 0.37 m/sec.
- Gas hold-up increased from 5.6% to 6.4%

"Spouting" Point:

- Minimum rpm required to achieve good bubble distribution at a given gas flow
- Defined by observation as lack of large coalesced bubbles reaching the surface
- ~80 rpm @ 15 Nm^3/hr air flow
- Found to be ~70 rpm @ design air flow during water runs in the commercial reactor

Mixing Time

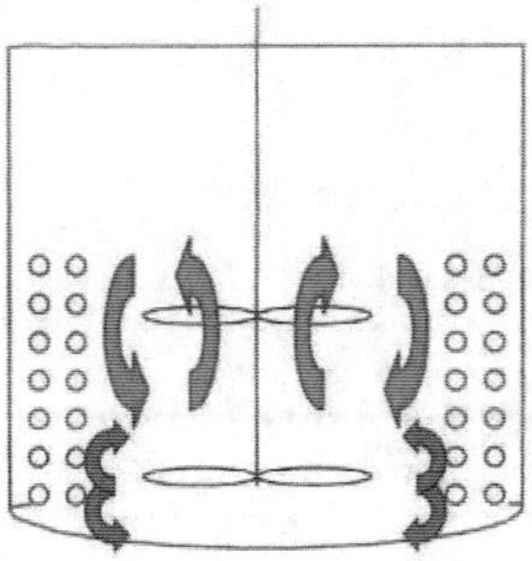

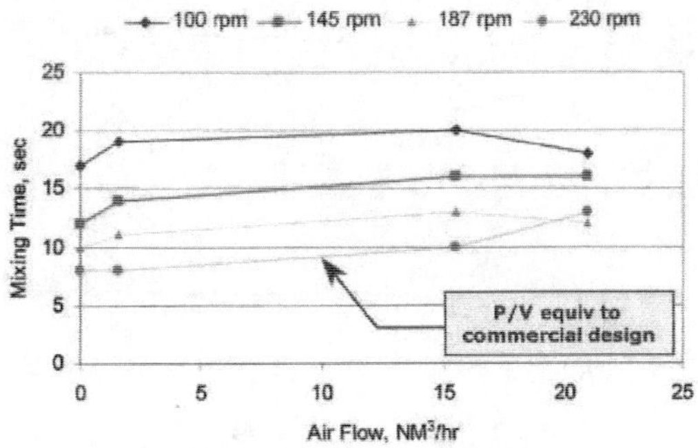

- Radial flow zone
  - intense gas-liquid contact for bubble formation
  - some finite residence time prior to the axial zone
  - minimizes headspace O2 – minimized gas by-passing
- Axial flow zone
  - enough energy input to maintain good bubble dispersion

~ 7 seconds of bottom zone segregation at design conditions

SABIC Innovative Plastics™

سابك sabic

Heat Transfer

Heat transfer scale-up critical - rate limiting during 1st stage oxidation

Jacket area/volume of commercial reactor ~40% of the pilot scale

$$Nu = C \cdot Re^{2/3} \cdot Pr^{1/3} \cdot (\eta/\eta_w)^{0.14}$$

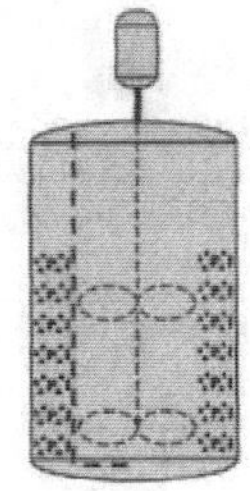

Generally for turbine type impellers:
C ~ 0.62 - for transfer to the jacket
up to 1.5 - for a single helical coil in the available literature

Issues/Questions:

- **Process side film coefficient = f(impeller type, impeller/coil geometry)**
- **Presence of coils reduce the film coefficient of the jacket**
- **Influence of multiple coil banks?**
- **Influence of air flow?**

Measurements with the proposed geometry were needed to reduce the risk of heat transfer scale-up

SABIC Innovative Plastics™

سابك sabic

Heat Transfer

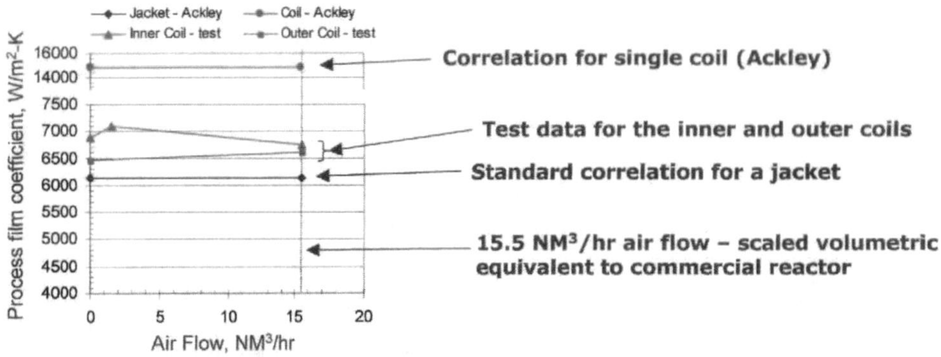

- Tests conducted in water - 25ºC to 35ºC
- C ~ 0.68 for the dual coil bank @ design air flow
- No significant effect of air flow seen
- Jacket area was ignored (used for design safety factor)
- Test data confirmed the adequacy of the dual coil design

Post Start-Up Plant Performance: +25% margin

SABIC Innovative Plastics™

سابك sabic

Mass Transfer

Reaction piloted with a single stage flat blade turbine (Rushton impeller)

- Need at least equivalent mass transfer (k_La) for the commercial reactors

$$k_La = A \cdot \varepsilon^B \cdot v_{sg}{}^C$$

where:
k_La is liquid mass transfer coefficient, hr^{-1}
ε is energy dissipation density, W/kg
V_{sg} is superficial gas velocity, m/hr

- Agitation tests in water comparing k_La for Rushton and proposed commercial impeller design

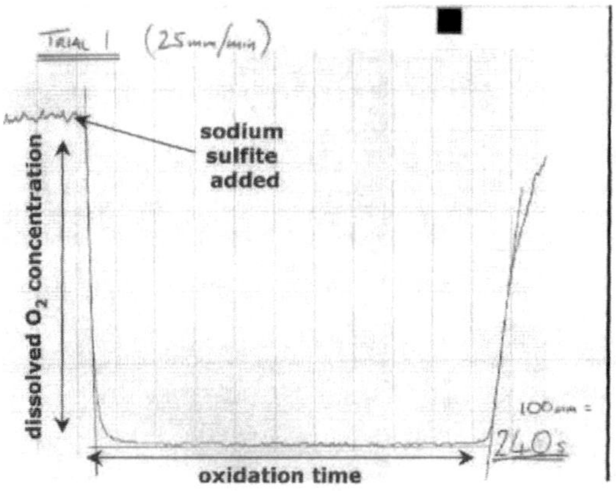

$$k_La = \frac{0.5 \cdot m \cdot MW_{O_2}}{t \cdot V \cdot c^* \cdot MW_{Na_2SO_3}}$$

where:
m mass of sodium nitrite added, g
MW molecular weight, g/gmole
T oxidation time, hours
V liquid volume, m^3
c* saturated O2 concentration, g/m^3

SABIC Innovative Plastics™

سابك sabic

Mass Transfer

Pilot scale reaction development:

- Single Rushton impeller
- P/V = 0.56 kW/m³

GOOD PERFORMANCE –
no sign of mass transfer limitation

Agitation trials @ 1m³ scale:

- Single Rushton vs. dual (proprietary) impellers at the same d/D
- In water at ambient T, P=atmospheric
- Scaled volumetric air flow

Higher k_La with proposed design

Proposed Design Shows Improved Mass Transfer vs. Pilot Reactor

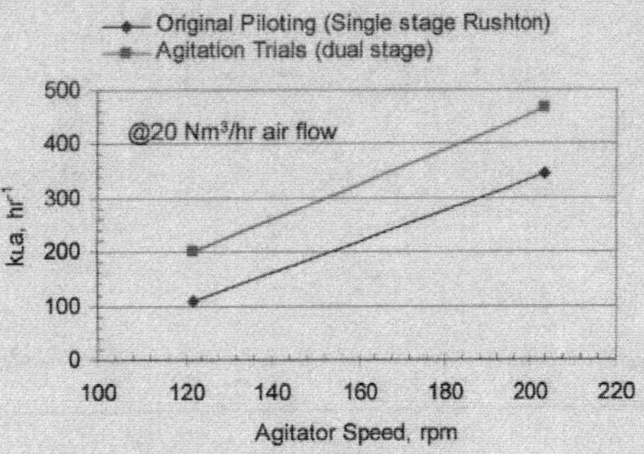

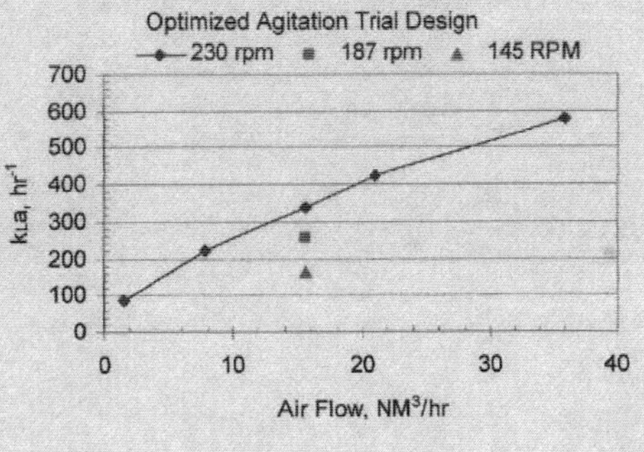

SABIC
innovative
Plastics™

Power Draw

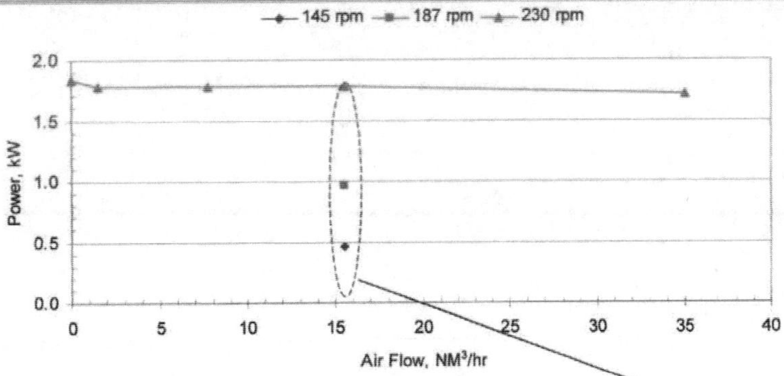

- *Power input ~ constant with changing air flow*

  Power/volume related to:
  - bubble size (original and coalesced)
  - bubble dispersion
  - gas hold-up

- Verified power proportional to $(\text{rpm})^{2.9}$
  ~ same as standard $(\text{rpm})^{3.0}$

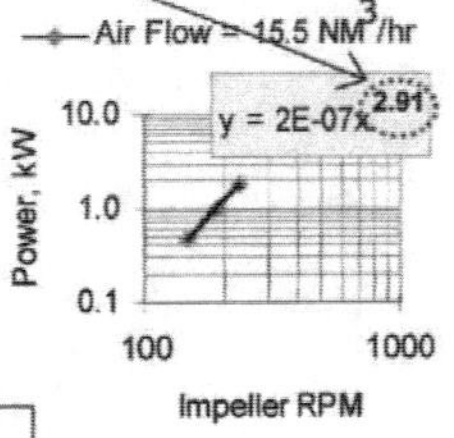

Improvement vs. Conventional Impellers - Constant P/V with Changing Gas Flow

SABIC
innovative
Plastics™

Summary

Goal Achieved – Successful Scale-Up

- Design basis cycle time demonstrated
- 25% heat transfer margin realized – room for cycle time reduction
- Mass transfer is not limiting
- Low headspace concentration indicates efficient use of oxygen
- Product quality duplicates the pilot plant results

SABIC Innovative Plastics™

سابك
sabic

(389c) Industrial Reactor Evaluation with Uncertainties In Mixing Performance and Reaction Kinetics

Michael D. Cloeter[1], Pramod Patil[1] and Aaron Sarafinas[2], (1)The Dow Chemical Company, Freeport, TX, (2)The Dow Chemical Company, Spring House, PA

Frequently the Fluid Mechanics & Mixing Discipline of The DowChemical Company (Dow) is asked to assess the mixing-sensitivity of industrial chemistries. The chemistries may be undergoing scale-up or are in some way experiencing processdifficulties. Sometimes the aim is to minimize impurities of any sort. Inother, perhaps more challenging, cases the aim is to maintain a constantdistribution of by-products to ensure that the product performance isunchanged. Most desirably, we have access to a detailed intrinsic kinetic ratemodel and detailed knowledge of the fluid dynamics, so that the pertinent timescales can be calculated and a Damkohler analysis completed for mixingsensitivity. Reality is frequently quite different.

Generally, we have the following scenarios with respect to kinetics:

- A kinetic model exists or can be derived from the correct type ofexisting lab or process data.

- A kinetic model does not exist, but a reactor mole balance & productdistribution is available or can be derived from process data. Reactorsampling may be impossible, meaning the engineer must resort to approximationsbased on plant feed and product flows, compositions, tank level changes, etc. Calorimetricdata may exist. If resources allow, a kinetic model can be derived from molecularmodeling and thermochemical kinetics, then validated with product data.

- No intrinsic kinetic data exist. Previous process data was taken with atransport limitation.

And generallywe have these scenarios for flow field:

- Vessel geometry is known precisely and the flow field iswell-characterized.

- Vessel geometry is known but flow characteristics are not known. If thereaction is localized, process instrumentation such as thermowells helpunderstand some spatial reaction characteristics.

- Studies in lab glassware, such as a round-bottomed flask, have givensatisfactory performance. Blending is inherently rapid. No significant changesto the flow field have been studied.

Furthermore,there can of course be large interactions between the turbulence and reactions.

The nextscale often must fit in to an already-existing reactor somewhere. In thiscase, the stakeholders need assurance that the existing reactor can work, and whetherany changes must be made to the agitation, feed introduction, etc. to ensuresuccess. Sometimes the reactor is yet to be built, so that many degrees offreedom for design are available.

This talkwill focus on real examples of recent industrial problems encountered and themethodologies used to address them.

See more of this Session: Mixing In the Process Industries - How to Develop a Successful Process
See more of this Group/Topical: North American Mixing Forum

(389d) Designing In Reliability

Todd M. Hutchinson, Research and Development, Philadelphia Mixing Solutions, Ltd, Palmyra, PA

Most rotating equipment is selected as part of a Capitol Project govern by rules from Procurement with specifications written by in house engineers or outsources engineering partners. Because of Project spending limits the equipment supplier who is believed to meet the specification requirements with the lowest bid receives the contract. There are times when the equipment supplier who meets the project costs targets are awarded the orders regardless of the specification requirements. And then there are those times when the specifications do not adequately describe a process resulting in improperly designed equipment.

Regardless of this traditional process used unfortunately those responsible for Reliability inherit equipment and are tasked to operate the equipment at peak performance for continually longer periods of time. Of course this is nothing new. The question is how does someone in charge of predictable operating a plant impact the Project Process and selection of equipment. Specific applications will be studied showing failed agitation equipments and the efforts over the years to improve reliability. Unfortunately these efforts were focused on symptoms and not the True Root Cause. In some cases the plants were designed based on the perfect word and did not account for upset conditions. In other cases the basic data for the design was flawed. Either way those responsible for reliability face a seemingly never ending battle. The studied applications include staggering reduction in operating costs that have led to companies changing their Project Process.

See more of this Session: Mixing In the Process Industries - How to Develop a Successful Process
See more of this Group/Topical: North American Mixing Forum

(389e) The Evolution of Process Understanding Through Micromixing and Mass Transfer Studies In the Successful Development of a Specialty Chemical Process

Aaron Sarafinas, Core R&D, Engineering Sciences, Fluid Mechanics & Mixing, The Dow Chemical Company, Spring House, PA

Researchers who develop new specialty chemical products are often faced with a variety of potentially conflicting rate processes which can influence reaction selectivity and final product properties. As the researchers' understanding of the process develops through laboratory studies and pilot plant trials, the original process paradigms must be critically examined in order to assure successful commercial implementation.

This paper examines the evolution of process understanding in the development of a specialty chemical product. The critical reaction step could produce the desired product (R) and two by-products (Q and S) by the reactions:

$$A + B \rightarrow Q$$

$$A + B \rightarrow R$$

$$R + B \rightarrow S$$

Researchers needed to minimize potential losses of the very expensive reagent B and the moderately expensive reagent A in the pilot plant. Recognizing the series-parallel nature of the R and S reactions, researchers used the Bourne reactions in the laboratory and found a strong correlation between the selectivity of the Bourne reactions and the production of R and S in the critical reaction step over a wide range of laboratory process conditions. The Bourne reactions were used in the pilot plant to establish operating conditions for the critical reaction step. The desired selectivity of the critical reaction step was achieved in the initial scale-up with high overall yield on the expensive reagents.

Though the initial results were rather satisfying, certain observations on the initial and subsequent scale-up campaigns indicated that the process understanding was not complete. This paper will further examine those observations and look in more detail at the phases, components, and rate processes controlling the process. Consideration of both the mass transfer effects and micromixing led to more complete process understanding and a highly successful commercial start-up.

(389f) Further Studies of Micromixing: Scale-up, Baffling and Feed Pipe Backmixing

Melissa Assirelli, Huntsman Polyurethanes, Everberg, Belgium and **Alvin Nienow**, Birmingham University, Birmingham, UK

The work reported was undertaken at the Institute of Chemical and Engineering Sciences, Singapore. It consisted of an experimental study of micromixing in a semi-batch reactor agitated by a Rushton turbine using two different reaction schemes. With one, the iodide-iodate scheme, at the same mean specific energy dissipation rate and feed location, the selectivity was similar at the 3.6 L scale to that found earlier at the 25 L scale (Assirelli *et al.*, 2002). Also, at the 3.6 L scale, a study of back mixing at different feed locations showed that it was greatest when feeding into the trailing vortex, which is also the best feed location to improve micromixing. At that location, all pipes gave a similar selectivity at the same exit velocity. However, on closer inspection, it was found that even at the same velocity, in general, the bigger the pipe, the greater the level of backmixing. This finding raises certain questions about micromixing based on local specific energy dissipation rates. However, it is in agreement with some previous work (Bourne *et al.*, 1981). Comparison with the limited literature also showed that one of the two criteria for predicting backmixing fitted the present results quite well (Vicum *et al.*, 2004) but the other overpredicted it by two orders of magnitude (Jo *et al.*, 1994). The other reaction scheme involved two parallel reactions, consisting of an acid-base neutralization and the alkaline hydrolysis of ethyl monochloroacetate (Bourne and Yu, 1994). It was used to compare micromixing in a 3L vessel under baffled conditions with that without them. Such a comparison has only been made once before at a larger 25 L scale and with the iodide-iodate scheme (Assirelli *et al.*, 2008). The earlier surprising result was confirmed, namely that the mean specific energy dissipation rate required for equivalent selectivity without baffles was less than that required with them.

REFERENCES

Assirelli, M., W. Bujalski, A. Eaglesham and A. W. Nienow, "Study of Micromixing in a Stirred Tank using a Rushton Turbine: Comparison of Feed Positions and Other Mixing Devices", *Chem. Eng. Res. Des.*, **80**, 855-863 (2002).

Assirelli, M., W. Bujalski, A. Eaglesham and A. W. Nienow, "Macro- and Micro-Mixing Studies in an Unbaffled Vessel Agitated by a Rushton Turbine", *Chem. Eng. Sci.*, **63**, 35– 46 (2008).

Bourne, J. R., F. Kozicki, U. Moergelit and P. Rys, Mixing and Fast Chemical Reaction - Model-Experiment Comparisons, *Chem. Eng. Sci*, **36**, 1655-1663 (1981).

Bourne, J. R. and S. Yu, Investigation of Micromixing in Stirred Tank Reactors using Parallel Reactions, *Ind. Eng. Chem. Res.*, **33**, 41-55 (1994).

Jo, M. C., W. R. Penney and J. B. Fasano, Backmixing into Reactor Feed Pipes Caused by Turbulence in an Agitated Vessel. *AIChE Symp. Ser.*, **90**, 41-49 (1994)

Vicum, L., S. Ottiger, M. Mazzotti, L. Makowski and J. Baldyga, Multi-Scale Modelling of a Reactive Mixing Process in a Semi-Batch Stirred Tank, *Chem. Eng. Sci.*, **59**, 1767-1781 (2004).

See more of this Session: Mixing In the Process Industries - How to Develop a Successful Process
See more of this Group/Topical: North American Mixing Forum

(467a) Modeling the Residence Time Distribution In a Batch Fermentor: Comparison of CFD Prediction with Experiment

Abhishek Dutta[1], **Rita Lencastre Fernandes**[2], Anna-Lena Heins[3], Anna Eliasson Lantz[3], Anker D. Jensen[2], Krist V. Gernaey[2] and Ingmar Nopens[1], (1)Biomath, University of Ghent, Ghent, Belgium, (2) Department of Chemical and Biochemical Engineering, Technical University of Denmark, Kgs. Lyngby, Denmark, (3)Department of Systems Biology, Technical University of Denmark, Kgs. Lyngby, Denmark

Continuousstirred tank reactors (CSTR) are ubiquitously used as fermenters in the biotechnologicalproduction of enzymes, pharmaceuticals, metabolites, antibodies, etc. They arealso used in the bioprocess industry for the production of bio-ethanol, bio-hydrogenand even in mammalian cell culture. In most cases, a scale-up of these processesis necessary to yield commercial success and viability. As with most scaling-upoperations, mixing is also a crucial parameter in a biological process and isoften not ideal when operating at industrial scale. Poor mixing in a CSTR may becaused by gas by-passing and back-mixing phenomena leading to the formation of deadzones where, in the case of a bioreactor, cells will experience a limited supplyof nutrients and oxygen. In large-scale bioreactors gradients of e.g.substrate, oxygen and pH arise as a result of non-ideal mixing, resulting in a spatiallyheterogeneous extracellular environment. This fact has been suggested to be themain reason behind lower product yields, but higher cell viability, in comparisonto lab scale cultivation (Enfors et al., 2001)

Understanding bioreactor hydrodynamics in general, andnon-ideal mixing in particular, is a key problem to understand the inherent bioprocess.The Residence Time Distribution (RTD) is one such way to characterize thenon-ideal mixing in a CSTR (Choi et al., 2004). Toestimate the RTD, a glucose tracer study has been performed experimentally in aSartorius CPlus batch bioreactor, havinga working capacity of 10 litres, and the results werecompared using Computational Fluid Dynamics (CFD) simulations. Several caseswith varying impeller speed have been considered to study the effect of mixingphenomena on the RTD. The mean and variance of the distribution are analyzed forthe extent of the deviation of the reactor behavior from its ideal state.Furthermore, the sensitivity of the sliding mesh is tested by varying theposition of the interface zone. This would eventually contribute to building anoptimized model of the bioreactor, and will be used for integration withbiological models aiming at improving performance prediction in the cases ofpoorly-mixed reactors.

Reference

[1]Enfors, S.O., Jahic, M., Rozkov, A., Xu, B., Hecker, M., Jrgen, B. et al. (2001). Physiologicalresponses to mixing in large scale bioreactors. J. Biotechnol., 85, 175-185.

[2] Choi, B.S., Wan, B., Philyaw, S.,Dhanasekharan, K., Ring, T.A. (2004). Residence Time Distribution in a stirredtank: comparison of CFD predictions with experiment. Ind. Eng. Chem. Res., 43,6548-6556.

See more of this Session: Mixing In the Production of Biopharmaceuticals
See more of this Group/Topical: North American Mixing Forum

(467b) Mixing It up: Probing Cell Culture Bioreactors Using CFD Modeling

E. Todd Sorensen[1], Pratik Jaluria[1], Luke Munholand[2] and Derek Adams[1], (1)Upstream Development, Alexion Pharmaceuticals, Cheshire, CT, (2)Fluent Inc., Lebanon, NH

Process scale-up in the biopharmaceutical industry is complex and empirical, because of the presence of living cells used to manufacture the desired product. Experimentation is increasingly difficult as the scale increases, due to regulatory and cost constraints, so there is a heavy reliance on scaling calculations and small-scale experiments. Simplified mixing experiments are often conducted in conjunction with standard bulk calculations to derive physical characteristics of each scale, such as volume-to-volume-per-minute (VVM), Power to Volume (P/V), Tip Speed, mass transfer coefficient (k_La), maximum shear stress, and recirculation time.

We scaled up one of our processes from the 2L to the 10,000L bioreactor initially by maintaining the VVM and P/V ratios, which resulted in a 30% drop in Maximum Viable Cell Density (VCD) with a 35% drop in productivity from the 2L to the 10,000L scale. Since we have chosen the 2L bioreactor to be our main development scale we wanted to increase our understanding about the differences between the 2L and 10,000L scales. In order to assist us, we are beginning to use Computational Fluid Dynamics (CFD) as an additional tool, allowing us to explore the internal fluid dynamics within each tank with more spatial and temporal detail. Our relatively simple CFD models are analyzed against directly-comparable physical experiments (where possible) to validate the results.

Thus far, we have determined that mixing times of liquids at large scale were 8 times longer and exhibited relatively long-lived regions of high concentrations of added components. This result is important for our processes, where significant quantities of non-physiological solutions (e.g., concentrated nutrient feeds and base) are routinely added to the reactor. We believe that the slower mixing in the large-scale bioreactors permits zones of different concentrations of components ("mini – environments") to exist when liquids are added to the bioreactor. These "mini-environments" certainly contain pH gradients, may exhibit hypoxic characteristics, likely have altered Carbon Dioxide levels, and/or contain elevated Osmotic environments. This indicates that scaling biological processes may require more than just maintaining bulk physical characteristics but instead rely on scaling mini-environments to allow for adequate gas exchange and liquid-liquid mixing. Overall, this work highlights our initial phase of development to enhance biological process scale-up using customizable CFD software.

See more of this Session: Mixing In the Production of Biopharmaceuticals
See more of this Group/Topical: North American Mixing Forum

EFFECT OF DIFFERENT AGITATION SYSTEMS AND PROTOCOLS ON THE GROWTH OF *NICOTIANA TABACUM* CELLS: COMPARISON OF GROWTH KINETICS IN ERLENMEYER FLASKS, CONVENTIONAL STIRRED TANKS, AND ECCENTRICALLY AGITATED TANK BIOREACTORS

Josefina Castillo-Reyna[1] and Mario M. Alvarez[2]

[1] *Departamento de Ingeniería Química y Biotecnología, Tecnológico de Monterrey at Estado de México, Estado de México, México*

[2] *Centro de Biotecnología-FEMSA, Tecnológico de Monterrey at Monterrey, Monterrey, NL. México*

Erlenmeyer flasks and stirred tank bioreactors are the most commonly used bioreactors for the lab culture of plant cells. We compare the performance of these systems against and novel eccentrically agitated stirred tank bioreactor for the particular case of culture of *Nicotiana tabacum* cells. We measured the specific growth rate, total cell density, cell viability and integrity/morphology over time of tobacco cells in 300 mL Erlenmeyer flasks, and 1L conventional and eccentrically stirred bioreactors. We observed that, under some agitation regimes, concentrically agitated systems inflicted damage to the tobacco cells, particularly at high cell densities. At these conditions, higher specific growth rates and cell densities were observed in Erlenmeyer flasks and eccentrically agitated bioreactors. Our results suggest eccentric bioreactors offer with a more benign and scalable growth environment for the culture of shear sensitive plant cells.

Key words: tobacco cells, stirred tank bioreactor, plant cell culture, mechanical stress, mass transfer

INTRODUCTION

The culture of plant cells in bioreactors has been suggested as a promising platform to produce high added-value bioproducts such as recombinant enzymes or specialized biopharmaceuticals. Plant cell culture requires of relatively inexpensive culture media, particularly as compared to animal cell culture. In addition, molecular techniques to transform plant cells, for example *Nicotiana tabacum*, have achieved a good degree of maturity: for example, the expression of molecules of high value in health (Magnuson et al., 1998; James et al., 2000; Hong et al., 2002; Judge et al., 2004). At laboratory scale, plant cells are routinely cultured in Erlenmeyer flasks. At an intermediate scale, airlift reactors for low to moderate cell densities, and vessels with paddle–type or helical-ribbon impellers for high cell density systems have been suggested as reactors for plant cell culture (Doran, 1999; Doran, 2000).

Research work on the area of reactor engineering for plant cells is limited. In general, suspension cell culture techniques in bioreactors are restricted to Erlenmeyer flasks, conventional stirred tanks, and air-lift bioreactors. In particular, limited kinetic data is available on the culture of tobacco cells in instrumented bioreactors.

A number of reviews have dealt with the choice of bioreactors for plant cell suspension culture (Doran *et al.*, 1993) and most emphasis has been placed on modifications of the conventional STR, with bubble aeration, employing a variety of impeller designs. However, equipment normally employed for microbial cultivation may not be immediately applicable to plant cells, due to differences between the characteristics of both individual cells and suspensions of the respective systems (Kieran *et al.*, 1997) that have direct implications for the design and scale up of suspension culture systems. Plant cells are large, and when they are exposed to turbulent shear fields where the eddy size approaches the cell size, the cells can be exposed to a twisting motion that can damage them. Lower levels of shear appear to affect cell surface receptors and nutrient transport. Reactors with high shear must be avoided. Stirred tanks designed for the culture of bacteria are not good choices, but modified stirred tanks can be suitable (Paul *et al*., 2004).

Here, we study the effect of three different agitation conditions on the specific growth rate, mortality rate, morphology, and viability of *Nicotiana tabacum* cells cultured in concentrically and eccentrically agitated instrumented bioreactors (1 L effective volume). Growth results from Erlenmeyer flask culture are also presented for comparison.

MATERIALS AND METHODS

Biological Material and culture media

Tobacco cells *Nicotiana tabacum* 1 (NT-1) were used in this project. Cells were obtained from *calli* kindly donated by Dr. Paula G. Miller from the Chemical and Biomolecular Engineering Department at Cornell University, Ithaca, N.Y. Tobacco cell suspension culture and *calli* were maintained in **NT1** medium containing 4.3 g Murashig & Skoog basal salts, 30 g sucrose, 50 ml MES (stock 20x), 10 ml B1-Inositol (stock 100x), 3 ml Millers I (stock) and 2.2 ml 2,4-D (stock) in deionized water and adjust pH to 5.7 with KOH. For solid medium include 8 g of agar (Mayo *et al*., 2006).

NT1 Cell Growth and maintenance in solid media

Tobacco cell suspension culture and *calli* were maintained in **NT1** medium containing 4.3 g Murashig and Skoog basal salts, 30 g sucrose, 50 ml MES (stock 20x), 10 ml B1-Inositol (stock 100x), 3 ml Millers I (stock) and 2.2 ml 2,4-D (stock) in deionized water. The pH of the medium was adjusted to 5.7 with KOH. For solid medium, 8 g of agar were added (Mayo *et al*., 2006) to this basic formulation. *Calli* were grown in Petri dishes covered with aluminum foil, placed in the incubator to 25° C and kept in darkness (Loyola-Vargas & Vazquez-Flota, 2006). After three incubation weeks, a re-culture was done. Antibiotic solutions were prepared and sterilized by filtration. They should be added after autoclaving when the medium has cooled to 50 °C (Mayo *et al*., 2006).

Cell culture experiments in flasks

Culture experiments of tobacco cells were conducted in Erlenmeyer flasks under different inoculums concentrations (1 g/L and 5 g/L). Erlenmeyer flasks of 1 L of capacity with 500 ml NT1 medium were used. Flasks were inoculated by adding 1 g/L or 5 g/L of *callus* previously established in agar plates (Mayo *et al*., 2006). Erlenmeyer flasks were inoculated with 12-16 days old suspension cultures. All flasks were covered with aluminum foil to keep them in darkness. Flasks were placed in an orbital shaker at 150 rpm and a temperature from 25 to 27°C (similar than in Loyola and Vazquez, 2006). A medium change was performed weekly. Flasks were sampled for analytical measurements by taking a volume of 5-ml daily. Specific growth rates at the exponential phase, maximum cell densities, and mortality rates were calculated for each Erlenmeyer experiment.

Cell culture in bioreactors

Culture experiments were conducted in 350 mL instrumented concentric stirred tank bioreactors (from Infors®, Switzerland) under 100, 200, and 300 RPM. Initial cell concentrations of 5 g/L were used. Additional culture experiments were performed in a 2L instrumented bioreactor (from Applikon®, Netherlands). The reactor multiport stainless steel head plate was modified according to a design by

Alvarez *et al.* (2008) to allow for concentric and eccentric agitation. Temperature was set at 26°±1°C in all bioreactor experiments.

Evaluation of cell biomass, viability and integrity

Cell biomass was determined by Fresh Cell Weight and dry cell weight evaluation from well mixed samples taken at different culture times (see for example Loyola & Vazquez, 2006). Due to the small sample size, freeze drying was used to dry. Packed Volume was evaluated by taking 5 ml of sample in a 15-ml conic centrifuge tube. Viability determinations were made daily through a staining using Fluorescein Diacetate (FDA) and Propidium Iodide (PI) Techniques. A drop of sample was placed in a slide and observed in 10X at the microscope. Cell counting was done by triplicate in different fields. An average was obtained and the results were expressed in percentage of viability (Loyola & Vazquez, 2006). For FDA observations filter no. 46 carl zeiss was used. Propidium Iodide is an intercalating agent and a fluorescent molecule used to stain DNA.

Carbohidrate determination

Sucrose, Glucose and Fructose concentrations were monitored during cell culture by HPLC from 5 mL supernatant samples. Carbohydrate quantification was performed in HPLC Waters model W1515 with a refractive index detector model W2414, using a SGE EXIL AMINO column in an isocratic run. The chromatographic conditions were: flow of 1 mL/min; mobile phase of Acetonitrile-Water 80: 20; column temperature of 30°C; detector Temperature of 30°C; and injection Volumen: 20 μL. Calibration Curves were made at several concentrations: 2,4,6,8 and 10 mg/mL for each carbohydrate. The standards used for calibration curves were : D-(-)- fructose > 99% (Sigma CAS 57-48-7) ; D-(+)-glucose > 99.5% (Sigma CAS 50-99-7) ; and sucrose >99.5% (Fluka CAS 57-50-1).

RESULTS AND DISCUSSION

Growth kinetics of tobacco cells

The growth behaviour of tobacco cells during exponential phase, both in Erlenmeyer flasks or instrumented bioreactors can be easily characterized using simple first order models. Figure 1 present a typical growth curve based on total biomass density (v.gr. g of fresh cell weight (FCW/L) and g of dry cell weight (DCW/L)) from tobacco cells cultured in Erlenmeyer flasks.

The curve exhibit a typical behavior, with a relatively large lag phase, and an exponential growth period followed by a slow decay in total biomass. Specific growth rates at the exponential phase, maximum cell densities, and mortality rates were calculated for each experiment. For the exponential growth phase, a simple first order model with respect to biomass (v.gr. $d[X]/dt= \mu [X]$) fits the data with correlation coefficients greater than $r^2 = 0.95$. In this equation, [X] is the total biomass (the model holds for both wet and dry cell biomass data), t is time and μ is the specific cell growth rate. Similarly, for the biomass decay phase, a model of the type $d[X]/dt= - k_m [X]$ describes the observed behavior. Therefore, both exponential growth and cell death can be adequately and simply described by only one parameter, μ and k_m respectively.

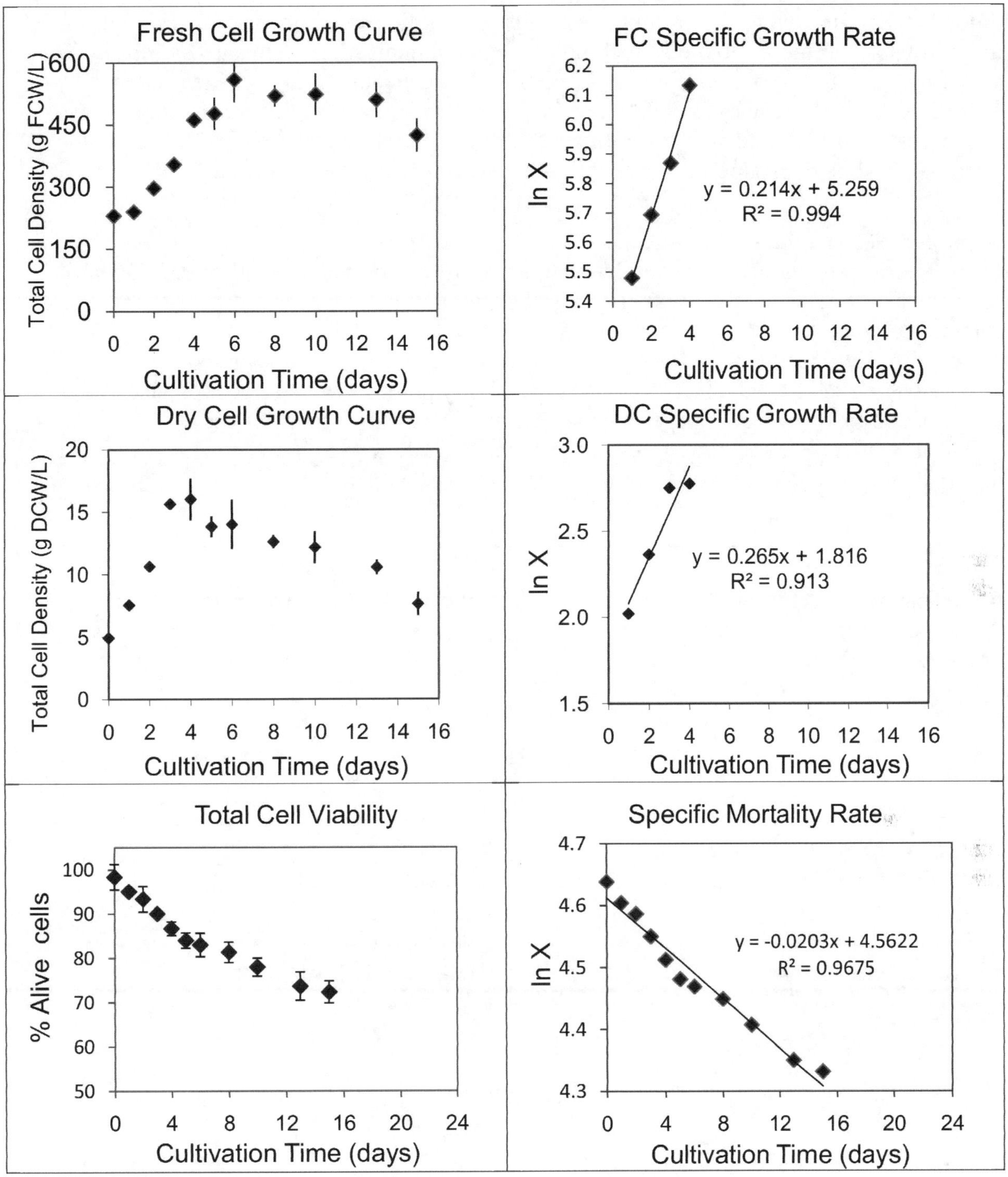

Figure 1. Growth curve of NT1 cells in 1 L flasks with an *inoculum* of 5 g dry cell/L (in orbital shaker at 26° C in darkness). Samples were taken every third day. Results corresponding to the average of 3 independent experiments are shown. Error bars indicate one standard deviation.

Growth of tobacco cells in Erlenmeyer flasks

We found that growth kinetic parameters do depend on inoculums concentration.In experiments conducted 1 g/L of inoculum, growth rates of tobacco cell fluctuate between 0.107 and 0.115 day^{-1}. In contrast, in cultures inoculated at 5 g/L, growth rates were estimated between 0.20 and 0.26 day^{-1}. Agitation conditions in Erlenmeyer flasks promote cell aggregation. Our experiments suggest that this aggregation, and probably a more efficient cell signalling/communication associated with it, is responsible from faster growth. We did not find evidence of cell fragmentation or damaged induced by the agitation conditions under Erlenmeyer agitation. In general, cell viability remained high (above 70%) for the first week of culture and was not influenced by initial cell concentration. Similarly, the mortality rate averaged 0.02 day^{-1} in experiments at the two different inoculums conditions.

Experiments in concentrically agitated bioreactors

Additional experiments in instrumented bioreactors were performed in 300 mL INFORS® bioreactors. The rational for these series of experiments was to simultaneously test different agitation conditions maintaining all the rest of the relevant operational variables the same. This experiment was of great importance to directly observe the effect of shear stress caused by high agitation speeds and measure its consequences on the growth kinetic parameters, cell morphology and aggregation, and yields of the culture. Figure 2 show results from experiments performed in Infors® bioreactors. DCW growth curves, (b) Total Cell Viability, (c) specific rates for growth, and (d) mortality results are presented. Runs at three different agitation values (100, 200 and 300 RPMs) are compared.

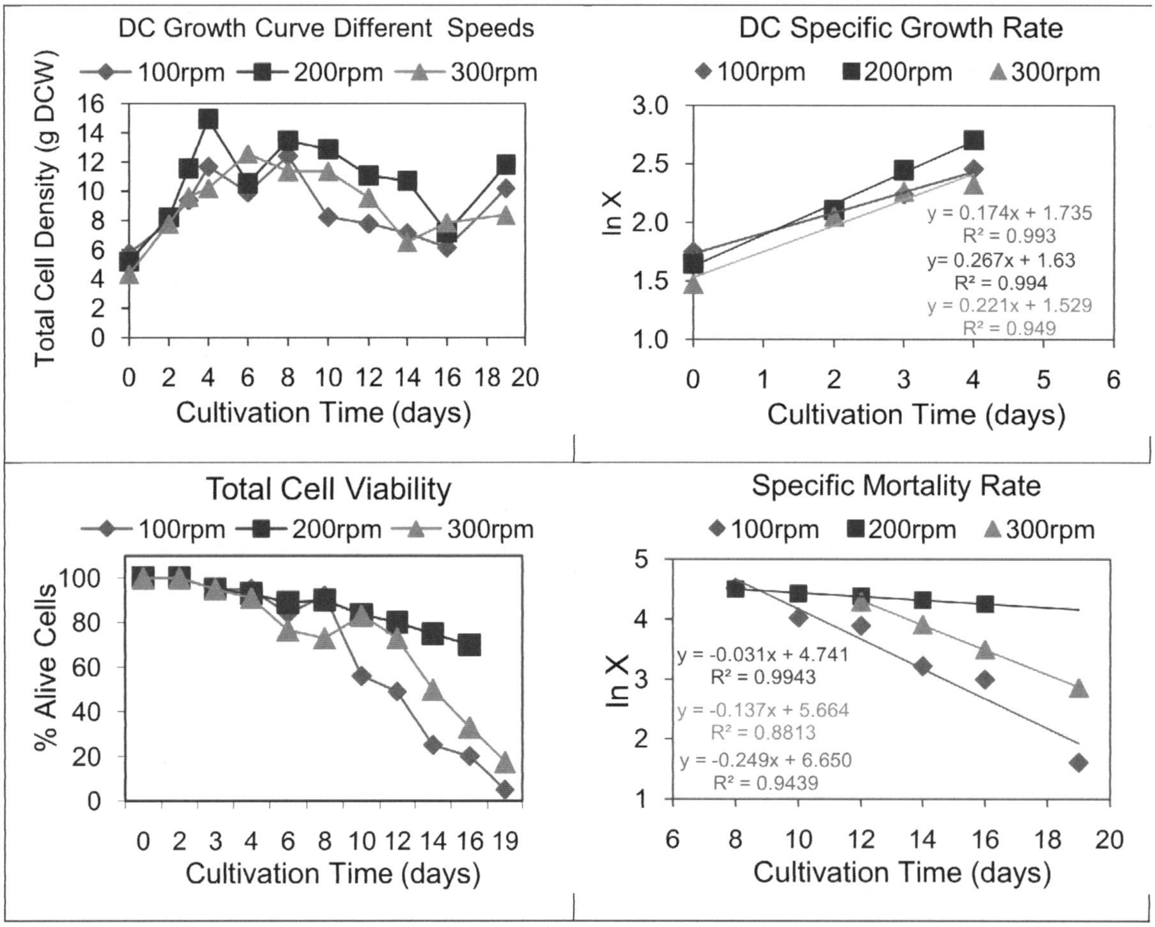

Figure 2. Growth kinetics of NT1 cells cultured in a 350 mL concentric bioreactor with NT1 medium at 3 different agitation speeds: 100, 200 and 300 rpm and 26° C in darkness. The initial *inoculum* was approximately 5 g dry cell /L.

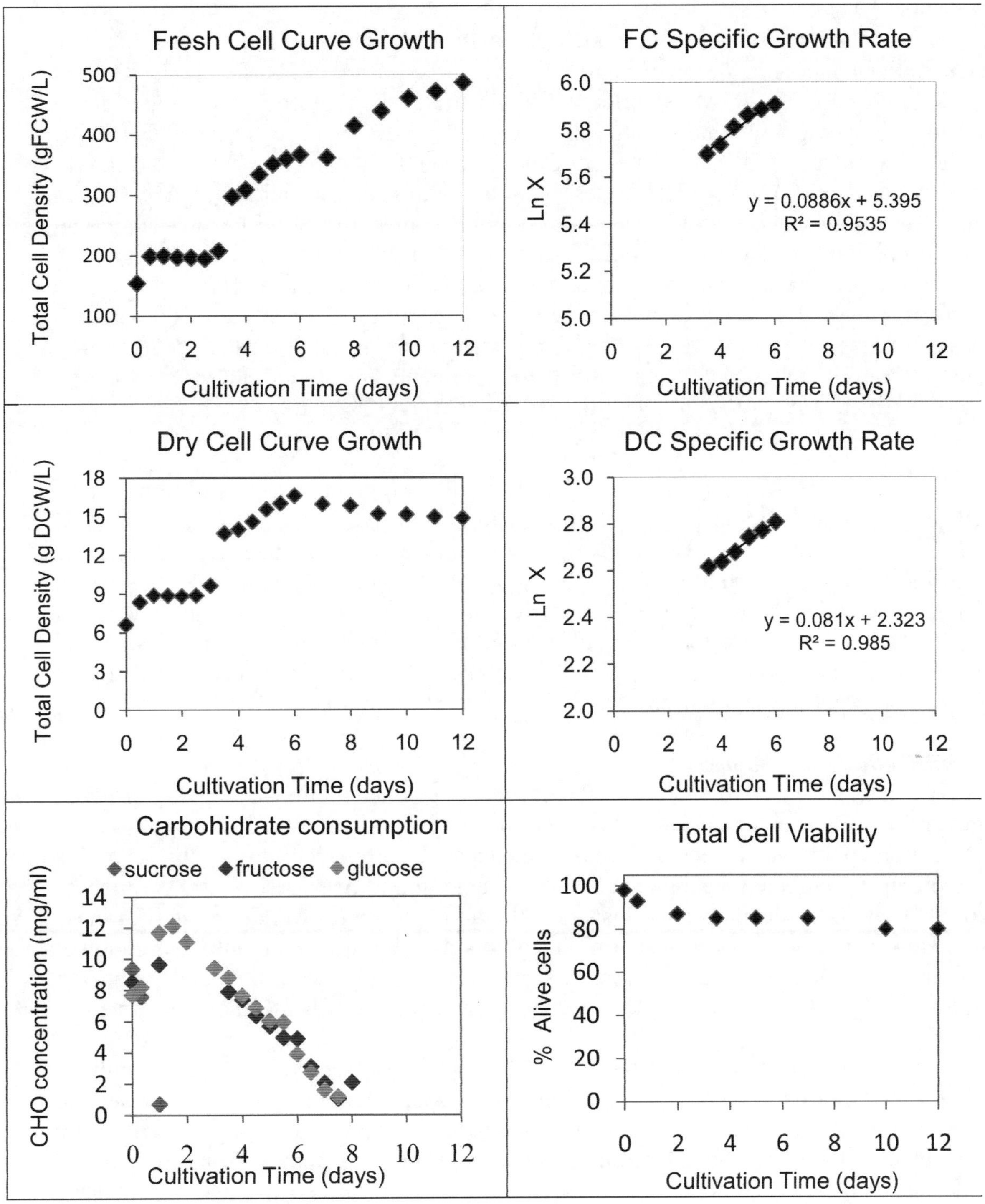

Figure 3. Growth curve (6) of NT1 cells in Applikon® bioreactor at 50 rpm for the first 4 days and 300 rpm until the end of the experiment, and 26° C in darkness. The *inoculum* was approximately 6.6 g dry cell /L.

Overall, better growth characteristics are observed at 200 RPM (i.e. a higher maximum cell density is achieved and higher cell counts are maintained throughout the evolution of the culture experiment. The specific growth rate observed at 100 RPM (μ =0.174 day^{-1}), is lower than those observed at higher speeds. The highest specific growth rate is observed when agitation is set at 200 RPM (μ=0.26 h^{-1}). This value is consistent to the highest specific growth rate observed in Erlenmeyer flasks at high inoculum concentrations. Not only the specific growth rate is maximized at 200 RPM, also the cell death rate is the lowest for the 200 RPM experiment, as illustrated by Figure 2. For experiments conducted at 100 rpm, the average maximum cell yield was 12.16 g DCW/L, an intermediate value when compared with the achieved at 200 and 300 rpm. The best yield obtained was 14.33 g DCW/L, corresponding to experiments at 200 rpm. In all cases, the maximum cell concentration was achieved between day 3 and 4. Regarding specific growth rates and doubling times, the highest growth rate (and therefore the minimum doubling time) was obtained in experiment at 200 rpm (0.294 days^{-1} and D_T= 2.35). The average specific growth rate during exponential phase for experiments at 100 and 300 rpm was practically identical, 0.23 and 0.24 days^{-1} respectively. In all experiments, cell viability remained above 80% for at least the first 8 days of culture. Remarkably, in experiments at 200 RPM, cell viability remained higher than 80% until day 14. In contrast, for the experiments at 100 and 300 rpm an abrupt decay in viability was observed after day 8, with viability values of 20 or 30% at day 19.

These results suggest that the most effective agitation speed to grow *N. tabacum* cell suspensions in Infors® bioreactors is 200 rpm. Results suggest that at 100 RPM, the culture growth is limited by oxygen mass transfer, and agitation intensity is not enough to sustain faster growth and higher cell densities. At 300 RPM, the agitation intensity is such that it induces cell damage, diminishing the proliferation rate and increasing the mortality rate. Vigorous agitation has been proposed as the reason for some plant cell lines to fail to grow optimally in stirred tank bioreactors. Plant cells have been regarded as being sensitive to shear stress because of their rigid cell wall, relatively large size and extensive vacuole volume. This hypothesis is supported by photographic evidence, taken at different cultivation times during the experiments at different agitation rates (not shown).

Experiments in eccentrically agitated bioreactors

Additional experiments were conducted at different *inoculum* concentrations and using different operational strategies in stirred tanks agitated with inclined disc impellers mounted in eccentric shafts. Eccentric bioreactor configurations have been referred as adequate for proper mixing at low Reynolds regimes, therefore avoiding cellular damage induced by intense agitation (Alvarez et al., 2002; Sánchez-Cervantes et al., 2006). Therefore, adequate cell densities and high growth rates were expected for these culture experiments. However, in experiments at low inoculums concentrations, namely 1 g/L, lower specific growth rate than in Erlenmeyer flasks were observed. Best results were obtained with high inoculums concentrations (5-6 g/L), with observed specific growth rates similar to those observed in Erlenmeyer flasks (Figure 3).

Two stage protocols in eccentrically agitated bioreactors were also assayed. In high inoculums experiments, cells were cultured under eccentric agitation at low speed, namely 50 RPM, for several days (during exponential growth) to later increase agitation speed to 300 RPM. Attending the scaling-up criteria of maximum tip speed velocity, this agitation condition would be comparable to more than 360 RPM in the 350 ml concentric bioreactors referred previously. However, no cell damage or increased mortality rate was observed in cell density high density cultures in eccentric bioreactors. Indeed, viabilities above 80% were sustained for twelve days of culture (Figure 3).

CONCLUSIONS

Our experimental results suggest than eccentrically agitated systems can be used as a cell culture system for shear sensitive plant cells. Similar specific growth rates were observed in highly inoculated eccentrically agitated bioreactors and Erlenmeyer flasks of similar scale (1-2 L). However, in smaller volume concentrically agitated bioreactors, mortality rates, cell viability and yields are a strong function of agitation rate. Overall yields comparable to those observed in Erlenmeyer flasks were observed only in a narrow window of agitation speeds, which complicates culture scalability.

REFERENCES

1. Alvarez, M. M., Arratia, P. E., & Muzzio, F. J. (2002). Laminar mixing in eccentric stirred tank systems. *Canadian Journal of Chemical Engineering, 80*(4), 546-557.
2. Doran, P. M. 2000. Foreign protein production in plant tissue cultures. Biotechnology. **11**:199–204.
3. Doran, P.M. 1993. Design of reactors for plant cells and organs. Advances in Biochemical Engineering Biotechnology. **48**: 117-168.
4. Doran, P.M. 1999. Design of mixing systems for Plant cell suspensions in stirred Reactors. Biotechnology Progress. **15**: 319-355.
5. Hong, S. Y., Kwon, T. H., Lee, J. H., Jang, Y. S. & Yang, M.S. 2002. Production of biologically active hG-CSF by transgenic plant cell suspension culture. Enzyme and Microbial Technology **30**: 763–767. Elsevier. South Korea.
6. James, E.A., Wang, C., Wang, Z., Reeves, R., Shin, J. H., Magnuson, N. S., & Lee, J. M. 2000. Production and characterization of biologically active human GM-CSF secreted by genetically modified plant cells. Protein Expression & Purification. **19**:131–138.
7. Joshi, J.B., Elias, C.B. & Patole, M.S., 1996. Role of hydrodynamic shear in the cultivation of animal, plant and microbial cells. Chemical Engineering Journal. **62**: 121–141.
8. Judge, N., A., Mason, H., & O'Brien, A. D. 2004. Plant Cell-Based Intimin Vaccine Given Orally to Mice Primed with Intimin Reduces Time of *Escherichia coli* O157:H7 Shedding in Feces. Infection and Immunity. **72:** 168-175. American Society for Microbiology.
9. Kieran P.M., MacLoughlin, P.F., & Malone D.M. 1997.Plant cell suspension cultures: some engineering considerations. Journal of Biotechnology. Vol. **59**: 39–52. Elsevier.
10. Loyola-Vargas, V.M. & Vázquez-Flota, F. 2006. Plant Cell Culture Protocols. Second Edition. Methods in Molecular Biology. Humana Press Inc.
11. Magnuson, N. S., Linzmaier, P.M., Reeves, R., An, G., Hay Glass, K., & Lee, J.M. 1998. Secretion of biologically active human interleukin-2 and interleukin-4 from genetically modified tobacco cells in suspension culture. Protein Expression & Purification. **13**:45–52.
12. Mayo, K.J., Gonzales, B.J. & Mason, H.S. 2006. Genetic transformation of tobacco NT1 cells with *Agrobacterium tumefaciens*. Nature Protocols. **1**: (3) 1105-1111.
13. Paul, L. E., Atiemo-Obeng, V.A.& Kresta, S.M. 2004. Handbook of Industrial Mixing: Science and Practice. John Wiley & Sons, Inc. Sponsored by the North American Mixing Forum.
14. Sanchez-Cervantes I. M., Lacombe J., Muzzio F.J., Alvarez M.M. 2006. Novel bioreactor design for the culture of suspended mammalian cells. Chemical Engineering Science. Volume 61, Issue 24, December 2006, pages 8075-8084. Part I: Mixing characterization.

(467d) Bioreactor Mass Transfer At Large Scale

Sperry Brown, Amgen, Inc., Thousand Oaks, CA

Mass transfer in bioreactors has been studied at large scale in Amgen's pilot plant. Implications of the relationships between sparging, agitation on mass transfer and reactor performance will be discussed.

See more of this Session: Mixing In the Production of Biopharmaceuticals
See more of this Group/Topical: North American Mixing Forum

(467e) Development of a Method to Measure Bioreactor Mass Transfer

Gregory Frank, Amgen, Thousand Oaks, CA

Determination of bioreactor mass transfer involves a number of decisions, including what conditions, media, experimental design and many other. The goal of developing a protocol is to have a rapid, low cost, and robust methodology that provides mass transfer data that's representative of conditions experienced in actual cell culture.

See more of this Session: Mixing In the Production of Biopharmaceuticals
See more of this Group/Topical: North American Mixing Forum

Experimental measurement and CFD modelling of bubble size distribution in an aerated stirred tank

ZbynŘk Kálal, Milan Jahoda, Institute of Chemical Technology, Prague, Czech Republic

Summary

The main topic of this study is experimental measurement and modelling of bubble size distribution in an aerated stirred vessel using CFD (Computational Fluid Dynamics) method. The studied system consisted of a cylindrical vessel of diameter $T = 0.29$ m, which was equipped with six-bladed Rushton turbine and filled with tap water to the height $H = T$. The secondary phase was introduced through a ring sparger situated under the impeller. Geometry and mesh of the computational domain were created using ANSYS DesignModeler and ANSYS Meshing software, calculations were carried out in CFD solver CFX 12.1 using Euler-Euler multiphase model together with k- turbulent model. The rotating impeller was modelled via MRF (Multiple Reference Frames) method and coalescence and breakage of bubbles using homogeneous MUSIG (Multiple Size Group) model with 13 classes of bubble size. Regarding the fact that these processes of particle size change represent from physical point of view very complicated and to a large extent stochastic phenomena (similarly as the very turbulent flow), the applicability of population balance models is presently very limited and numerical results should be always compared and fitted against experimental data.

Introduction

Presently the one-phase systems are well explored both experimentally and using simulations. Detailed description of dynamic behaviour of multiphase systems and determination of hydrodynamical variables is a difficult task because of complexity of interaction between phases. The presence of dispersed phase influences the dynamic of disperging phase. Interactions between phases include various mechanisms of momentum exchange like drag, lift and virtual mass force. However, in stirred tanks only contribution of drag force is significant and the others can be neglected during calculations [1].

The accuracy of results is usually affected by the particle size distribution modelling of the secondary phase, whether it is solid particles, droplets or bubbles. The size change can be caused by combination of different phenomena such as nucleation, growth, dissolution, aggregation (coalescence) or breakup of the particle. To capture this process it is necessary to use additional equation describing changes in particle size distribution together with transport or enthalpic equations for multiphase flow. This equation is generally referred to as population balance equation (PBE) and is used to describe phenomena such as crystallization, precipitation, granulation, fluid polymerization or even coalescence and breakage of bubbles in bubble columns and stirred tanks.

To simplify and speed up the calculation, constant size of dispersed (gas) phase is often assumed during mathematical modelling of liquid-gas system. This idea, however, does not fully capture the real situation, since via bubble-bubble and bubble-liquid interactions coalescence and breakage occur. So the constant bubble size model cannot correctly describe interfacial forces and other interactions between phases as heat and mass transfer, because their magnitude is dependent just on the bubble size.

The aim of this text is to give basic theoretical description of particle coalescence and breakup and possibilities of their modelling using CFD method in an aerated stirred vessel.

Theoretical part

Population balance equation and its solution

In general, PBE is a balance equation of the number density probability of some particle property (usually size). It defines how population of particles develop over time. Neglecting growth and nucleation term the general PBE can be written as follows

$$\frac{\partial}{\partial t}[n(V,t)]+\nabla\cdot[\vec{U}n(V,t)]=B_{ag}(n,V,t)-D_{ag}(n,V,t)+B_{br}(n,V,t)-D_{br}(n,V,t) \quad (1)$$

where $n(V,t)$ is the bubble number density function. It is defined as the number of bubbles with volume between V and $V+dV$ per unit volume and per class width dV. Thus, integration of $n(V,t)$ over all bubble volumes results in the total number of bubbles per volume. $B_{ag}(n,V,t)$, $B_{br}(n,V,t)$ express birth of bubbles of size V to $V+dV$ due to coalescence and breakage, $D_{ag}(n,V,t)$, $D_{br}(n,V,t)$ corresponding particle death, see following equations

$$B_{ag}(n,V,t)=\frac{1}{2}\int_0^V a(V-V',V')n(V-V',t)n(V',t)dV' \quad (2)$$

$$D_{ag}(n,V,t)=\int_0^\infty a(V,V')n(V,t)n(V',t)dV' \quad (3)$$

$$B_{br}(n,V,t)=\int_V^\infty g(V')\beta(V\,|\,V')n(V',t)dV' \quad (4)$$

$$D_{br}(n,V,t)=g(V)n(V,t) \quad (5)$$

Here $a(V-V',V')$ is the coalescence rate between bubbles of size $V-V'$ and V', $g(V')$ breakup rate (frequency) of bubbles of size V', i.e. the fraction of particles of volume V' breaking per unit time, and $\beta(V\,|\,V')$ probability density function of particles breaking from size V' to a particle of size V.

From the former equations it is obvious that PBE is a complicated integro-partial differential equation. As a consequence, it is possible to solve it analytically only in rare cases. Therefore various numerical methods are used to deal with it, e.g. Monte Carlo method, Method of Classes - CM, Standard Method of Moments - SMM or Quadratic Method of Moments - QMOM. Presently the best explored method is the CM, which is based on discretizing of continuous size range of the particle population in a finite number of size intervals (usually called bins). Each of these intervals is represented by one specific value and each particle can interact with particles in other groups. The PBE for class i can be written as

$$\frac{\partial}{\partial t}(\rho_g n_i)+\nabla\cdot(\rho_g\vec{U}_g n_i)=\rho_g(B_{ag,i}-D_{ag,i}+B_{br,i}-D_{br,i}) \quad (6)$$

where n_i represents the number of bubbles of given class per unit volume

$$n_i=\int_{V_i}^{V_{i+1}} n(V,t)dV \quad (7)$$

As mentioned earlier, every class is represented by a specific value of particle size V_i. However, due to coalescence and breakage phenomena also particles, the volume of which V lies somewhere between the values of two neighbouring classes V_i and V_{i+1}, can be produced. This problem is solved by splitting the particle into two neighbouring classes. More information about the CM including detailed derivation and discretization of source terms can be found in [2].

Method of Classes is used especially in cases when the size of largest and smallest particle are known and this range is not too big, so that it is possible to discretize the population with relatively small number of intervals. To gain sufficiently exact results it is recommended to use at least 12-15 classes. The main advantage of this attitude is direct computation of particle size distribution, although at the expense of significant computational demands, because one transport equation has to be solved for every bubble class.

Coalescence and breakup models

The modelling of bubble, drop or solid particle coalescence and breakup is very complex and it is based on the knowledge of coalescence and breakup rates and daughter size distribution (see equations 2-5).

Coalescence between bubbles occurs in three steps. Firstly, the bubbles collide together and trap a small amount of liquid between them. Subsequently, this liquid drains till the liquid film reaches its critical thickness. Finally, the film ruptures and bubbles aggregate. In reality, the duration of collisions is limited and coalescence will occur only if the interaction time is sufficiently long enough for the film to drain out to critical thickness, therefore the concept of collision efficiency is introduced. The resulting coalescence rate is then determined as a product of collision frequency and collision efficiency.

Various mechanisms of coalescence are known presently. Buoyancy-driven collisions result from the difference in rise velocities of bubbles having different sizes and thus different buoyancy force acting on them. Furthermore, during the rise of a bubble through a liquid some of this liquid is carried up and accelerated behind this bubble, which is known as a wake. When another bubble enters the wake region of the leading bubble, it will accelerate and may coalesce with the preceding one. However, all these mechanisms are usually neglected, and in the stirred tanks with high level of turbulence especially, in comparison with turbulent-induced coalescence, which result from random motion of bubbles due to turbulent fluctuations.

The random motion of fluid particles in a turbulent flow is usually assumed to be similar to the random movement of gas molecules, thus most of the present models for coalescence were derived analogously to classical kinetic theory of gases, where collisions between molecules are considered. Thus, various coalescence models show similar trends, e.g. a monotonous increase in the coalescence rate with increase in the particle diameter.

One of the most widely used coalescence model is the one given by Prince and Blanch [3]. The collision frequency ω_{ag} and collision efficiency P_{ag} are expressed as following

$$\omega_{ag}(V_i, V_j) = 0.356\pi (d_i + d_j)^2 \varepsilon^{1/3} (d_i^{2/3} + d_j^{2/3})^{1/2} \tag{8}$$

$$P_{ag}(V_i, V_j) = \exp\left(-\frac{\rho_l^{1/2} r_{ij}^{5/6} \varepsilon^{1/3}}{4\sigma^{1/2}} \ln \frac{h_0}{h_f} \right) \tag{9}$$

where equivalent radius r_{ij} is

$$r_{ij} = \left(\frac{1}{2} \left(\frac{1}{r_i} + \frac{1}{r_j} \right) \right)^{-1} \tag{10}$$

and h_0, h_f are initial and critical liquid film thickness separating the two coalescing particles (10^{-4} a 10^{-8} m).

Bubble breakage generally depends on the balance between external stresses that disrupt the bubble, and viscous stresses, represented by surface tension σ, that resist the bubble deformation. Most of the published breakage models are based on similar assumptions: 1) breakup is caused by turbulent eddies of size λ hitting the surface of bubbles with diameter d; 2) for the breakup to occur, the eddies must have energy higher than the surface energy of the bubble and simultaneously they should be smaller or equal to the size of the bubble; 3) eddies bigger than the bubble are supposed just to transport this bubble. Following these arguments, breakup frequency is given, similarly as in the case of coalescence, by a collision rate between particles and turbulent eddies multiplied by a collision (or breakage) efficiency. Because of the second and third assumptions, the breakage rate of particles should decrease with their decreasing size. For very small bubbles the breakup rate is close to zero, because the eddies which are even smaller than the bubbles do not have enough energy to break them and larger eddies only transport these bubbles.

The main difference among the available models is in their prediction of daughter size distribution (DSD), which represents the fraction of bubbles of size V' that break into bubbles of the size V. Some of the models assume a uniform distribution [3]. Others has a shape of truncated normal distribution which is centered at the half of the particle size [4]. It means, that these models prefer equal-sized breakage. In contrast, some other models favours unequal breakup [5], i.e. a division of the mother particle in one small and one big daughter particle, which is motivated by the fact that the unequal-sized breakup is less energetically demanding than the equal-sized one and thus should be more probable.

Luo and Svendsen based their model on assumption of isotropic turbulence and considered only binary breakup. The significant difference between this model and its predecesssors is that it gives not only overall breakage rate, but also partial breakage rate, i.e. the breakage rate for a particle of size V' splitting into a particle of size V and its complement. This partial breakup rate is expressed as

$$g(V,V') = C_B 0.9238 \frac{\varepsilon^{1/3}}{d^{2/3}}(1-\alpha_g)\int_{\xi_{\min}}^{1} \frac{(1+\xi)^2}{\xi^{11/3}} \exp\left(-\frac{12 c_f \sigma}{\beta \rho_l \varepsilon^{2/3} d^{5/3} \xi^{11/3}}\right) d\xi \tag{11}$$

where σ is the surface tension between phases, ξ dimensionless eddy size, c_f change in the surface area, β is a constant equal to 2.05 and C_B a calibration coefficient.

$$c_f = f^{2/3} + (1-f)^{2/3} - 1, \quad f = \frac{V}{V'}, \quad \xi_{\min} = 11.4\frac{\eta}{d}, \quad \eta = \left[\frac{\left(\frac{\mu_l}{\rho_l}\right)^3}{\varepsilon}\right]^{1/4} \tag{12}$$

where η is Kolmogorov micro-scale. The total breakup frequency of particles of size V' is then computed as an integral over all possible daughter particle sizes.

$$g(V') = \frac{1}{2}\int_0^1 g(V,V')df = \int_0^{0.5} g(V,V')df \tag{13}$$

One half before the integral is introduced because during integration every breakup is included actually twice. The integral is symmetrical around the value 0.5, so that a change of the upper limit of integration to this value and removing the one half from in front of the integral would yield the same result. A great plus of this model is that the DSD is derived directly from the model for breakage rate and found by normalizing the partial breakage rate by the overall breakage rate.

$$\beta(V \mid V') = \frac{1}{V'}\frac{g(V,V')}{g(V')} = \frac{2\int_{\xi_{\min}}^{1} \frac{(1+\xi)^2}{\xi^{11/3}} \exp\left(-\frac{12 c_f \sigma}{\beta \rho_l \varepsilon^{2/3} d^{5/3} \xi^{11/3}}\right) d\xi}{V' \int_0^1 \int_{\xi_{\min}}^{1} \frac{(1+\xi)^2}{\xi^{11/3}} \exp\left(-\frac{12 c_f \sigma}{\beta \rho_l \varepsilon^{2/3} d^{5/3} \xi^{11/3}}\right) d\xi df} \quad (14)$$

Experimental part

Experiments were carried out in a cylindrical vessel with inner diameter $T = 0.29$ m, which was filled with tap water to a height $H = T$ (Fig. 1). Four baffles of width $b = T/10$ and thickness of 10 mm were placed evenly around the perimeter of vessel wall. The tank was equipped with a standard six-bladed Rushton turbine of diameter $D = T/3$, the distance of which from the bottom was also $C = T/3$. It was mounted on a central shaft rotating with constant speed of 300 rpm. Halfway between the vessel bottom and lower edge of the impeller there was located a ring sparger, on the upper side of which there was six point outlets with a diameter of 1.5 mm through which air was introduced in the system. Volumetric gas flow rate was equal to 0.2 vvm.

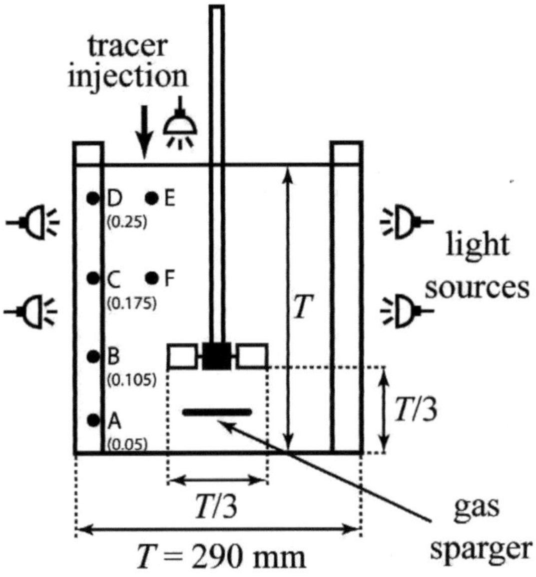

Fig. 1: Geometry of the stirred tank with six points for bubble size distribution measurement (A-F); the numbers in parentheses refer to vertical distance from the vessel bottom [m]

The determination of bubble size distribution (BSD) was accomplished using photographic analysis. Firstly, the tank was placed in a rectangular, glass vessel also filled with water, which prevented the distortion of the view when looking through rounded wall. The photographing and consequent evaluating of dense dispersions is generally very difficult due to overlapping of bubbles. Therefore, the tank was placed in a darkened room and a set of lights was placed on both sides of the tank and above the free liquid surface. These lights enlightened just the plane where bubble snapshots were taken, so that the unwanted background was significantly suppressed and identification of bubbles situated in the plane was facilitated.

The BSD was investigated at six points (see Fig. 1), which were all located in the vertical plane between baffles. Positions A-D were placed in the vicinity of the wall (approximately in the distance of two centimeters), points E and F then halfway between vessel wall and impeller shaft. During taking photographs a needle of known dimensions was placed in the desired position and served as a point at which the image was focused and simultaneously as a scale in the process of evaluating of photographs. Since the fluid moves in the mixed tank at considerable speed, it was necessary to use a low value of the exposure time, 1/1250 s, to gain sharp contours of the bubbles. Furthermore, appropriate depth of field was adjusted so as to obtain sharp outlines only of those bubbles that were found in the plane of interest. At each point about 50 photographs was taken. The captured bubbles were manually identified and their size was evaluated via image analysis in NIS-Elements software. The equivalent diameter of a spherical bubble was determined from the area of the circle which was set equal to the area of the bubble evaluated using image analysis.

For the simulations it was also needed to get the size of bubbles at the gas inlet because this parameter represents one of the boundary conditions of the numerical model. However, even at low gas flow rates gas jetting occurs, which prevents the identification of bubble size. Therefore, the flow rate was reduced to as low a value that at gas sparger separated bubbles were formed. Their size was used as input parameter of the model.

Regarding the fact that local gas hold-up measurement is very complicated and requires sophisticated equipment such as CT [6], only global gas hold-up was measured on the principle of liquid height elevation. Because the liquid surface in the tank fluctuate during stirring and so does not allow simple reading of the elevation, a sequence of photographs of the surface was acquired under ungassed and gassed condition. Using image analysis the averaged position of the liquid surface was determined for each condition and according to the difference in water level the gas hold-up was calculated.

Computational part

Geometry of the studied system was created in the preprocessor DesignModeler. The domain was classically divided into rotating part containing impeller and stationary part. For mesh generating another preprocessor was used Meshing. The final mesh contained 185 000 cells (Fig. 2).

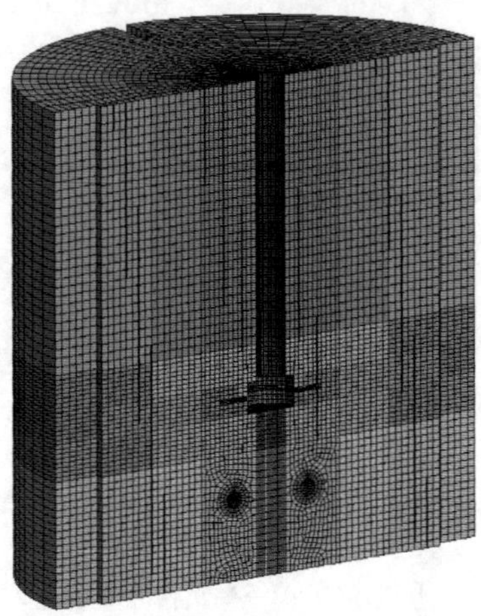

Fig. 2: Computational mesh

CFD simulations were carried out in software CFX 12.1. For numerical modelling of multiphase flow the most complex Eulerian model was chosen, so that for each phase a set of transport equations is solved. Because the system was in fully turbulent regime ($\mathrm{Re} = 4.49 \cdot 10^4$), *k*- turbulent model was used. The impeller was modelled via the MRF method. The upper part of the domain was set as a degassing boundary, which prevents water to leave the domain, but allows air to escape. The system was considered as isothermal and due to negligible solubility of air components transport between phases was omitted. On the basis of experiments the bubble diameter at the inlet was set to 8 mm. The PBE was solved using CM with 13 bubble groups in range from 0.1 to 16 mm, so that thirteen additional equations was necessary to solve. Coalescence kernel was calculated using Prince & Blanch model, breakage term then using Luo s model. The drag force was approximated according to the correlation given by Ishii and Zuber [7], other interaction forces were neglected.

Results

Experiments

At each investigated point 250-500 bubbles was marked and evaluated. The maximum bubble size observed in the system was around 6 mm, whereas the smallest detectable bubbles were of 0.1 mm in diameter. Most probably smaller bubbles existed, but due to limitations of the camera resolution and human eye during manual evaluation of taken photographs these were not incorporated. However, these bubbles contain only a very low fraction of gas volume and interfacial area. More than 90 % of both gas volume and interfacial area is stored in bubbles bigger than 1 mm. The average diameter varied from 0.5 to 1.2 mm and around 60 % of detected bubbles had diameter smaller than 1 mm.

To view bubble size distribution a quantity called Sauter mean diameter (SMD) d_{32} is usually used, which characterizes an average particle size. It is especially important in calculations where the active surface area is crucial.

$$d_{32} = \frac{\sum_i n_i d_i^3}{\sum_i n_i d_i^2} \tag{15}$$

The lowest SMD was detected in the impeller discharge stream where the turbulent dissipation energy has the largest value, which causes intensive breakup of bubbles (Fig. 3). The bubble diameter then slowly increases from the impeller towards the liquid surface and from the vessel wall towards the shaft because in the volume above impeller plane coalescence prevails. The smallest SMD was observed in position A. This was caused by the fact that the system was working in loading regime when vast majority of bubbles follows the upper recirculation loops and only very small bubbles are captured by the lower loops. These are then trapped here, coalesce and slip upward along the wall. On the contrary, if the system works in complete dispersion regime, the SMD in this position is considerably higher than in position B, because large bubbles are formed from bigger bubbles entrapped in the lower recirculation loops. The average diameter of bubbles entering the system was determined to be equal to 8 mm.

The gas hold-up in the system was found to be equal to 0.7 vol.%, which corresponds well with the correlation for systems with one Rushton turbine given by van't Riet [8].

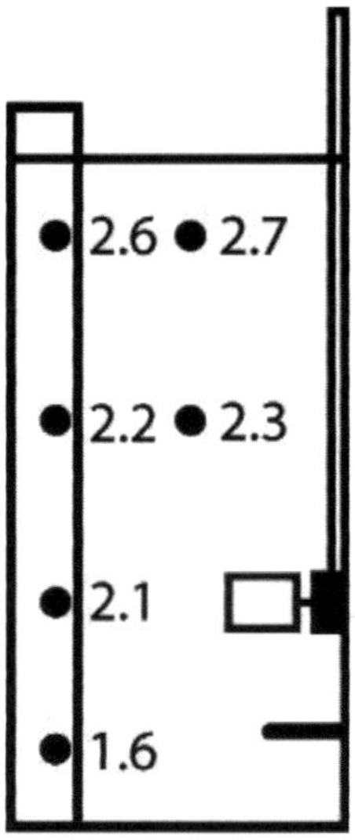

Fig. 3: Measured SMD in the tank [mm]

CFD simulations

As mentioned earlier, the bubbles were divided in 13 classes ranging from 0.1 to 16 mm in diameter. The breakup model by Luo was used. However, this model showed very strong underestimation of the SMD in the domain, especially in the impeller discharge stream. Practically all bubbles here were completely smashed in very small particles with diameter close to the size of the smallest class, i.e. 0.1 mm. This is caused most probably due to U-shaped daughter size distribution of the model. In Fig. 4 there are depicted dimensionless DSD curves for three different turbulent dissipation rates and constant bubble diameter and gas volume fraction.

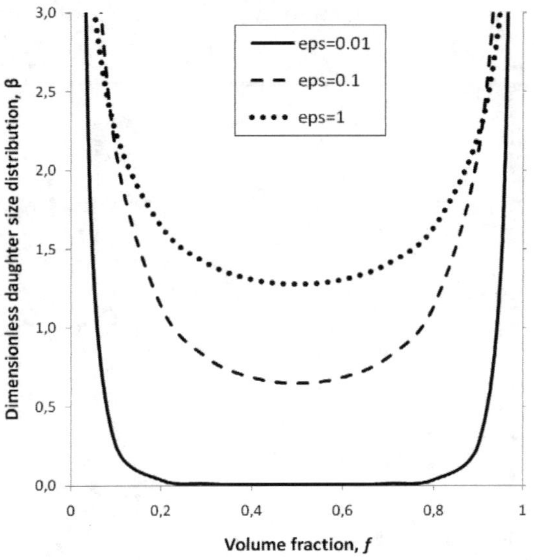

Fig. 4: Evolution of the dimensionless daughter size distribution of the Luo and Svendsen model with the change of turbulent dissipation energy (*d*=0.008 m, =0.01)

From the Fig. 4 it is obvious, that the model greatly favours the formation of a very small daughter particle and its complement, which stays after the breakup process nearly as big as the original mother particle. Considering the fact that the entering particles have diameter equal to 8 mm, thus being very unstable in turbulent flow of the system and having high breakup frequency, large number of small particles is with high probability gradually peeled from these bubbles. The breaking mother particle produces a lot of very small particles while still maintaining a high breakup rate. In the end, a great number of particles of the smallest size is created and the resulting SMD is underpredicted. The situation is depicted in Fig. 5.

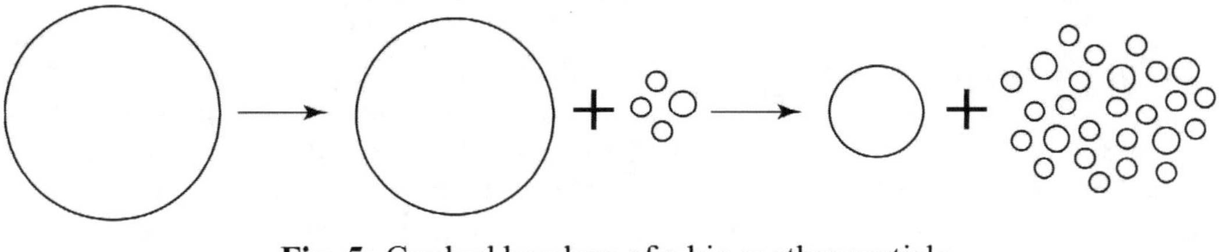

Fig. 5: Gradual breakup of a big mother particle

To remedy this issue either breakup rate must be decreased by adjusting an appropriate calibration coefficient of the breakup model, or the size of the smallest bubble group must be increased which results in formation of bigger bubbles by the breakup. However, even tenfold decrease of the breakup rate brought only a partial improvement of the resulting SMD which remained significantly lower than the experimental values (Fig. 6a). Much better agreement between simulations and experiments was achieved by setting the size of the smallest bubble class equal to 1.5 mm (see Fig. 6b). Nevertheless, this approach is not optimal either. Neglecting of particles smaller than is the size of the smallest class change the bubble size distribution because small bubbles are excluded. On the other hand, keeping in mind that a significant majority of bubble volume and interfacial area is stored in bigger bubbles, this method could be used as a good approximation for e.g. mass transfer simulations. A combination of adjusting the calibration coefficient of the breakup model and the size value of the lowest class seems to be the most powerful way how to handle the model given by Luo, but still it requires fitting against experimental data and its real predictivity is questionable. Use of different breakup models ([9],[10]) with DSD less emphasizing the formation of very small bubbles should be considered.

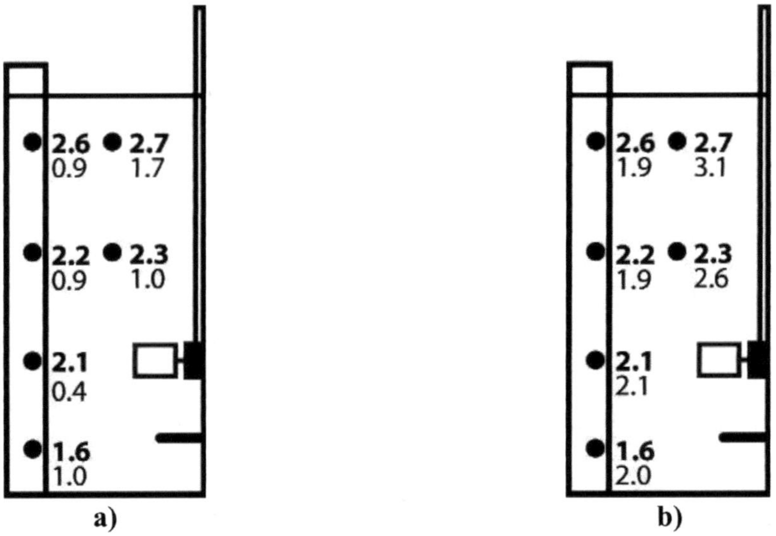

Fig. 6: Sauter mean diameter [mm] (1. line - experiments, 2. line - simulations)
a) The breakup rate equal to $0.1 \cdot g(V')$
b) The smallest bubble size $d_{b,\min} = 1.5$ mm

The gas hold-up predicted by Ishii-Zuber correlation was about 30 % lower than the value gained from the experiments. Therefore, turbulent drag correction should be included in the simulations. This correction is based on introducing of effective viscosity (sum of molecular viscosity and a term proportional to turbulent viscosity) in the computation of Reynolds number which enters calculation of the drag coefficient [11].

Conclusion

The bubble size distribution was measured in an air-water stirred tank using photographic analysis and these data were used for comparison with numerical results. For computations the breakup term by Luo and coalescence term by Prince and Blanch were used. However, this combination showed great underestimation of the predicted Sauter mean diameter. This is probably caused by the U-shaped daughter size distribution of the breakup model which significantly favours formation of very small particles, thus decreasing the value of the SMD. Therefore, fitting against experimental data is needed. This can be achieved by either decreasing of the breakup rate (alternatively increasing of coalescence rate) or adjusting of the bubble size of the smallest class. The latter attitude distorts the BSD because no very small bubbles (smaller than the lowest class) are formed, but since a great majority of bubble volume and interfacial area is stored in bigger bubbles, this method could be used as a good approximation for e.g. mass transfer simulations. Generally, the available breakup models are derived and validated only for certain flow conditions and their transferability is very limited, so that further research and development of these models is necessary.

Acknowledgements

This project has been supported by the Czech Science Foundation (Grant: 104/09/1290 and 104/08/H055) and by the Czech Ministry of Education (Grant: MSM 6046137306).

References

1. Scargiali F., D Orazio A., Grisafi F., Brucato A. (2007), "Modelling and simulation of gas-liquid hydrodynamics in mechanically stirred tanks", *Chemical Engineering Research and Design 85*, pp. 637-646
2. Kumar S., Ramkrishna D. (1996), "On the solution of population balance equations by discretization-I. A fixed pivot technique", *Chemical Engineering Science 51*, pp. 1311-1332
3. Prince M.J., Blanch H.W. (1990), "Bubble coalescence and break-up in air-sparged bubble columns", *AIChE Journal 36*, pp. 1485-1499
4. Martínez-Bazán C., Montanes J.L., Lasheras J.C. (1999), "On the breakup of an air bubble injected into fully developed turbulent flow. Part 1. Breakup frequency", *Journal of Fluid Mechanics 401*, pp. 157-182
5. Luo H., Svendsen H.F. (1996), "Theoretical model for drop and bubble breakup in turbulent dispersions", *AIChE Journal 42*, pp. 1225-1233
6. Khopkar A.R., Rammohan A.R., Ranade V.V., Dudukovic M.P. (2005), "Gas-liquid flow generated by a Rushton turbine in stirred vessel: CARPT/CT measurements and CFD simulations", *Chemical Engineering Science 60*, pp. 2215-2229
7. Ishii M., Zuber N. (1979), "Drag coefficient and relative velocity in bubbly, droplet or particulate flows", *AIChE Journal 25*, pp. 843-855
8. van't Riet K. (1975), "Turbine agitator hydrodynamics and dispersion performance", *PhD Thesis*, University of Delft
9. Lehr F., Millies M., Mewes D. (2002), "Bubble-size distributions and flow fields in bubble columns", *AIChE Journal 48*, pp. 2426-2443
10. Laakkonen M., Moilanen P., Alopaeus V., Aittamaa J. (2007), "Modelling local bubble size distributions in agitated vessels", *Chemical Engineering Science 62*, pp. 721-740
11. Bakker A., van den Akker H.E.A. (1994), "A computational model for the gas-liquid flow in stirred reactors", *Chemical Engineering Research and Design 72*, pp. 573-582

(546a) Experimental Validation of a Novel Model for the Micromixing Intensity

Frans Visscher, Xiaoping Chen, John van der Schaaf, Mart H.J.M. de Croon and Jaap C. Schouten, Laboratory of Chemical Reactor Engineering, Eindhoven University of Technology, Eindhoven, Netherlands

Experimentalvalidation of a novel model for the micromixing intensity in a rotor-statorspinning disc reactor

Frans Visscher, Xiaoping Chen, John van der Schaaf, Mart de Croon, and Jaap C. Schouten

Laboratory of Chemical ReactorEngineering, Eindhoven University of Technology, TheNetherlands

Introduction

Thecharacteristic timescale for micromixing are determined for a rotor-statorspinning disc reactor. This reactor consists of a rotating disc which isenclosed by two stationary discs and a cylindrical housing and is shown inFigure 1.

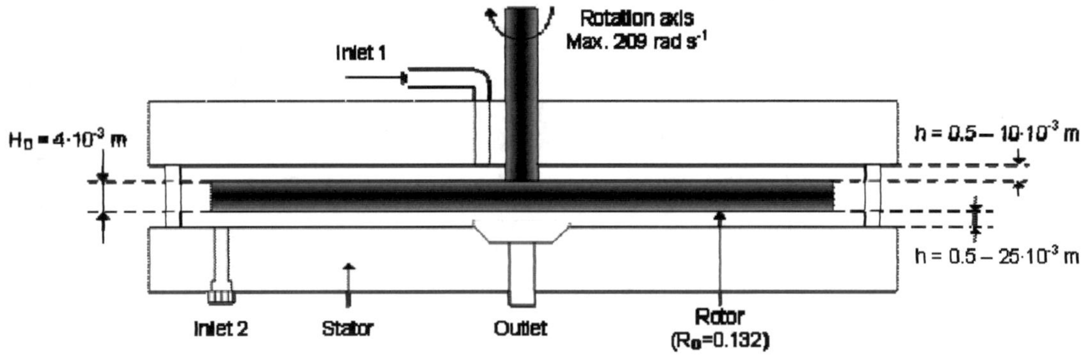

Figure 1. Rotor-stator spinning disc reactor configuration. Uponrotation of the rotor, the shear forces between the rotor and the stator causehigh turbulence intensity.

Previousresearch showed that intense mixing of multiphase processes can be achieved inthis reactor (Visscher, van der Schaaf et al. 2011). For a chemical processwith competing reactions, the yield and selectivity are influenced by themixing efficiency at micro scale in a reactor (Baldygaand Pohorecki, 1995). The mixing at micro scale canbe characterized with the iodide-iodate reaction system (Fournier, Falk et al.1996b;Fournier, Falk et al. 1996a). This reactionsystem is based on two parallel reactions that are competing for the acidicprotons.

Reaction 1:	$H_2BO_3^- + H^+ \quad H_3BO_3$	(pseudo instantaneous)
Reaction 2:	$5\,I^- + IO_3^- + 6\,H^+ \quad 3\,I_2 + 3\,H_2O$	(fast reaction)
Reaction 3:	$I^- + I_2 \quad I_3^-$	(pseudo instantaneous)

The Dushmanreaction (Reaction 2) is fast, in the same range of the micromixing process, butis much slower than the pseudo-instantaneous neutralization reaction (Reaction1). Therefore the product distribution of the reactions 1 and 2 can be used torepresent the mixing intensity at micro scale. This product distribution at thereactor outlet can be determined by measuring the triiodideconcentration with in-line UV-VIS spectroscopy.

The product distribution iscorrelated to the timescale of the micromixing process through a novel modelwhich is applicable for ideally mixed systems. This correlation is a functionof turbulence intensity, and operational conditions like the specific protondonor, volumetric flow rates, concentrations, and the reactor temperature. Thetrends predicted by the model are validated using the rotor stator spinning discreactor.

Reference List

Visscher,F., van der Schaaf, J., de Croon, M.H.J.M., and Schouten, J.C., 2011Liquid-liquid mass transfer in a rotor-stator spinning disc reactor. Annual AIChE meeting 2011, Minneapolis,MN.

Baldyga, J. and Pohorecki, R., 1995. Turbulentmicromixing in chemical reactors -- a review. The Chemical EngineeringJournal and the Biochemical Engineering Journal 58, 183-195.

Fournier,M.C., Falk, L., and Villermaux, J., 1996a. A new parallel competing reaction system for assessing micromixingefficiency - Experimental approach. ChemicalEngineering Science 51, 5053-5064.

Fournier,M.C., Falk, L., and Villermaux, J., 1996b. A new parallelcompeting reaction system for assessing micromixing efficiency--Determinationof micromixing time by a simple mixing model. ChemicalEngineering Science 51, 5187-5192.

See more of this Session: Novel Mixer and Mixed Reactor Design
See more of this Group/Topical: North American Mixing Forum

(546b) Challenges In Gas Dispersion Impeller Design

Wojciech Wyczalkowski[1], Robert Higbee[2] and Chris Hibshman[2], (1)Philadelphia Mixing Solutions LTD, Palmyra, PA, (2)R&D, Philadelphia Mixing Solutions LTD, Palmyra, PA

Gas dispersion impellers have been long used in the industry, such as Rushton, Smith, ICI Gas Foil (Middleton), PMSL GDX, and others with sharp concave blades. Often it is necessary to retrofit existing impellers with new impellers to accommodate process changes and/or energy savings.

The ability of adjusting impeller characteristics is very important while retrofitting the existing equipment. Often the rotational speed of mixer and its diameter cannot be changed therefore the only way to retrofit is to adjust impeller geometry and specifically Power Number. Geometry change also affects the gas/ungassed characteristics which can be determined by experiment only. The paper presents experimental data and analysis of family of gas dispersion impellers that is based on the original GDX Gasfoil (PMSL's GDX gas dispersion impeller) design. In addition, the structural strength and the cost also were considered.

Test impellers were made by the stereo lithography process that allows for precise production from 3D CAD models.

See more of this Session: Novel Mixer and Mixed Reactor Design
See more of this Group/Topical: North American Mixing Forum

(546c) Oil In Water Emulsions Using Confined Impinging Jet Mixer

Shad W. Siddiqui and Ian T. Norton, Chemical Engineering, University of Birmingham, Birmingham, United Kingdom

High value food-grade products often require creation of micron size and smaller emulsions which can be stored for long periods of time for consumption and other applications. Emulsification using impinging jets ensures fast droplet break-up under intense mixing and high energy dissipation conditions, apart from providing a continuous alternate to batch-wise emulsification process. Using food-grade sunflower oil as the dispersed phase and a combination of oil volume fraction, emulsifier type and concentration and hydrodynamic conditions small emulsions of 3-4μm have been made (Siddiqui et al., 2011). Variation of disperse phase (oil) viscosity and oil types is expected to have significant impact on the end emulsion droplet size distribution and is investigated in this work. Emulsification is carried out in presence in presence of Sodium dodecyl sulfate (SDS) and Tween 20 emulsifiers. The effect of repeated shear by re-circulating the emulsion through the impinging jet device is also studied. Multiple passes through the device allow subsequent breakage of droplets, besides allowing sufficient time for emulsifier to coat the newly formed droplet surface, thus avoiding re-coalescence.

Reference:

Siddiqui, S.W., and Norton, I.T. (2011) Investigation of the emulsification process of oil in water emulsions using confined impinging jets (Submitted)

See more of this Session: Novel Mixer and Mixed Reactor Design
See more of this Group/Topical: North American Mixing Forum

(546d) Axial-Flow High-Volume / Low-Head Pump Design Energy Method

Robert Higbee, Wojciech Wyczalkowski and Binxin Wu, R&D, Philadelphia Mixing Solutions Ltd., Palmyra, PA

It has been common practice to use classic pump curves when designing axial flow high-volume / low-head pumps. This presentation discusses axial pump design that is based on performance equations derived from an energy balance perspective. A specific real-world design case is presented for a pump with an output of 11,400 gallons per minute @ 6 foot head. Propeller 18.25 inch design OD was a required precondition. Primary design goal - maximize pumping efficiency. Leading edge prop geometry computed from incoming flow vector. Trailing edge prop geometry derived from energy-based performance equations with a slip-factor term borrowed from the world of marine prop design. Computational Fluid Dynamics (CFD) was used to predict the rotation-flow component of the fluid flow exiting the prop which was used to design the flow-straightener. A method for predicting a reasonable test RPM range will be discussed. A full-scale lab test was performed to pinpoint the final operating RPM within the predicted range and to verify the expected design performance. CAD models were made using 3-D parametric CAD from PTC (Wildfire™ 4.0). Performance of pump was simulated using the mathematic computation software Mathcad™. Part design strength was verified by Finite Element Analysis (FEA, Pro/Mechanica). Fluid flow was predicted by CFD software from Fluent™. Pump test conformed to Hydraulics Institute HI 2.6 and US Army Corps of Engineers standard EM 1110-2-3105. Pump test evaluation included DigitalFlow™ ultrasonic flow meter, pressure transducers, and strain gauge based shaft power measurement using Microstrain Inc's V-link™ wireless RF telemetry.

See more of this Session: Novel Mixer and Mixed Reactor Design
See more of this Group/Topical: North American Mixing Forum

(546e) FLUID DYNAMIC Analysis of FLOW IN Orbiting Dishes

Jonathan M.D. Thomas[1], Amlan Chakraborty[1], M. Keith Sharp[2] and **R. Eric Berson**[1], (1)Chemical Engineering, University of Louisville, Louisville, KY, (2)Mechanical Engineering, University of Louisville, Louisville, KY

It is well documented that hemodynamic parameters, especially wall shear stress (WSS), have been shown to play important roles in altering various endothelial cellular responses, intracellular pathways and gene expression, and to have significant impacts on disease progression, such as atherosclerotic plaque development. However, the fluid dynamics in commonly used orbiting cell culture dishes and the effects of the resulting oscillating fluid flow on shear and endothelial cellular responses within the dishes have not previously been accurately determined due to the difficulty in characterizing the complex flow. WSS in orbiting dishes is oscillatory rather than steady due to the travelling waveform and varies across the surface of the dish at any instant in time. The first part of this effort presents a cfd model which provides complete spatial and temporal resolution of oscillatory WSS over the bottom surface of an orbiting Petri dish throughout the orbital cycle. The model was reasonably well validated by the limited analytical solution.A dimensional analysis study was then performed to describe flow and shear characteristics as a function of Slope ratio, Froude Number, and Stokes Number, which allows for a significant reduction in the computational effort to characterize flow characteristics in orbiting dishes. Transition points for each parameter were determined for low and hi values of the remaining parameters. The contribution of tangential and radial WSS components across the bottom surface was then determined as a function of each parameter.

See more of this Session: Novel Mixer and Mixed Reactor Design
See more of this Group/Topical: North American Mixing Forum

(617a) A Modeling Strategy to Predict the Hydrodynamic In Stirred Tanks with Moving Free Surface

Mourad Heniche, Dept. of Chemical Engineering, cole Polytechnique de Montral, Montral, QC, Canada

The knowledge of the shape of the free surface in stirred tanks is of particular interest for the security of the equipments as well as for the quality of end-use products to name a few. The prediction of the free surface topography by numerical simulation is in general a difficult problem. Most CFD studies were conducted assuming a fixed flat free surface restricting the range of operating conditions where no free surface deformation should occur. To alleviate this limitation, we assume that the free surface moves along the axial direction of the vessel. The prediction of its configuration is achieved by the liquid level height function technique. This approach is implemented within the finite element method framework in combination with the fictitious domain method for the discretization of the governing equations of change as well as for the handling of the impeller. In the presentation, after a summary of the body of knowledge on free surface flows in agitated tanks, we will present the numerical methodology developed and the results obtained considering a common technology impeller for the mixing of a Newtonian fluid in laminar-transition regime.

See more of this Session: The Use of CFD In Simulation of Mixing Processes
See more of this Group/Topical: North American Mixing Forum

(617b) Predicting Two-Dimensional Bubbling Bed Flows Using Gas-Solid Continuum Modeling

Deepak Rangarajan, Chemical Engineering, University of Florida, Gainesville, FL and Jennifer Sinclair Curtis, University of Florida, Gainesville, FL

The dense phase gas-solid continuum equations are ridden with purely empirical and poorly validated constitutive relations which make one question the universality of their application to different flows. In this talk, a specific type of dense-phase flow with important industrial applications, namely, the spouted bed flow is studied in detail using the Department of Energy's gas-solid continuum code called MFIX. The performance of the different commonly used closure models in predicting both the macroscopic and the differential scale flow behavior is analyzed by comparing with experimental results available in literature. Some of the issues involved in ensuring accurate experimental comparisons, such as taking into consideration the influence of sidewalls present in the experimental setups, are explored. The effect of fluidizing the bed of particles prior to jet injection is also investigated.

See more of this Session: The Use of CFD In Simulation of Mixing Processes
See more of this Group/Topical: North American Mixing Forum

(617c) Effect of Agitation Speed On the Velocity Profiles and Shear Strain Rate Variability In the USP Dissolution Testing Apparatus 2

Yimin Wang, Ge Bai and **Piero M. Armenante**, Otto H. York Department of Chemical, Biological and Pharmaceutical Engineering, New Jersey Institute of Technology, Newark, NJ

The fluid velocity profiles at different locations inside a standard USP Dissolution Testing Apparatus 2 were experimentally obtained via Laser Doppler Velocimetry (LDV) at three impeller agitations speeds, namely 50 rpm, 75 rpm and 100 rpm. The experimental results were compared with the predictions obtained with Computational Fluid Dynamics (CFD) where the κ-ω model with low Reynolds number correction was used to account for turbulence effects. In general, good agreement was found between the experimental LDV velocity measurements and the CFD simulation predictions. The non-dimensional tangential, axial and radial velocity profiles (scaled with the impeller tip speed) and the flow pattern were found to be nearly independent of the agitation speed in most regions of the vessel, implying that increasing the agitation speed generally produced a corresponding increase in the local values of the velocity. However, the velocity profiles and flow pattern in the inner core region just below the impeller, where the dissolving tablet is usually located, were found to be much less sensitive to agitation speed. In this region, the axial and radial velocities were especially low and were not significantly affected by agitation increases. This inner core region at the center of the vessel bottom persisted irrespective of agitation intensity. The CFD predictions also indicated that increasing the agitation speed resulted in a higher shear strain rate distribution along the vessel bottom, although the strain rate was always very low at the center of the vessel bottom, even when the agitation speed was increased.

See more of this Session: The Use of CFD In Simulation of Mixing Processes
See more of this Group/Topical: North American Mixing Forum

(617d) CFD Modeling of Solid Suspension In a Stirred Tank for Pharmaceutical Application

Shitanshu Ratilal Gohel[1], **Shital Joshi**[1] and Mohammed Azhar[2], (1)Support and Services, Ansys Fluent India Pvt Ltd., Pune, India, (2)Tech Support/Consulting, Ansys UK Ltd, Sheffield, United Kingdom

Solid-liquid mixing is one of the most important mixing operations in pharmaceutical industry due to its vast applications in many unit operations such as crystallization, adsorption, dissolution and solid-catalyzed reaction. In this study, a computational fluid dynamics (CFD) model was developed using two fluid model for solid-liquid mixing process in a typical stirred tank reactor. The multiple reference frames (MRFs) approach was used to simulate impeller rotation in a weakly baffled reactor. The turbulent solid-liquid flow was simulated using a two-fluid Reynolds Stress Model with mixture properties. The effect of turbulent dispersion force is also taken in to account. Two different drag laws were tested. Simulation results were compared with various experimental data including cloud height and solid distribution. Parametric simulations are run at different rotational speed to calculate just suspension velocity.

See more of this Session: The Use of CFD In Simulation of Mixing Processes
See more of this Group/Topical: North American Mixing Forum

(617e) The Effect of Numerical Diffusion On Scalar Mixing Prediction Using CFD

Minye Liu, DuPont Engineering Technology, Wilmington, DE

Numerical diffusion is an intrinsic part of a numerical method for a convection-diffusion flow system. When CFD is used to predict mixing and reactions in a chemical reactor, the predicted results are often tainted by numerical diffusion since the physical diffusion, molecular or turbulent, could be overwhelmed by the numerical diffusion. This is almost guaranteed to be the case for scalar mixing in a laminar flow of industrial size. It has been believed that turbulent diffusion is several orders of magnitude larger than molecular diffusion, and therefore, turbulent mixing prediction should not be affected by numerical diffusion. In this talk, a method is first presented to estimate the order of magnitude of numerical diffusion in a CFD solution of a continuous flow reactor. Several examples, from microfluidic mixers to industrial sized stirred tank reactors, will be discussed and the magnitude of both physical and numerical diffusivities will be compared. Effect of numerical diffusion on mixing time predictions will also be discussed.

See more of this Session: The Use of CFD In Simulation of Mixing Processes
See more of this Group/Topical: North American Mixing Forum

(617f) Blending of Liquids Starting From a Stably Stratified State

Jos Derksen, Department of Chemical & Materials Engineering, University of Alberta, Edmonton, AB, Canada

Blending of liquids starting from a stably stratified state

We have conducted directnumerical simulations of blending of two miscible liquids with differentdensity and equal kinematicviscosity in a baffled mixing tank equipped with an axial impeller (a 45-degreespitched-blade turbine). As the starting point of each simulation the denserliquid occupies the lower half of the tank volume, the lighter liquid the upperhalf, and the flow velocity in the tank is zero everywhere. At time equals zerowe switch on the impeller and monitor the development of the flow and mixturefraction field in the tank. In addition to the geometrical layout of the tankand impeller, the flow system is governed by two dimensionless numbers: aReynolds number (Re) and a Richardson number (Ri). Reynolds numbers are in the range 3,000 to12,000 so that we cover transitional and mildly turbulent flow regimes and areable to fully resolve the flow (direct simulations). In the base-case theimpeller is down pumping and placed with an off-bottom clearance of one-thirdof the tank height. In this situation mixing patterns and associated mixing timesshow interesting trends with respect to the Richardson number. At low values ofRi mixing patterns are akinto what is observed for Ri=0 (left figure): theinterface between light and heavy liquid is sucked into the impeller swept volumeand macro-mixing occurs relatively quickly with somewhat longer mixing times(compared to the situation with Ri=0). If Ri ³

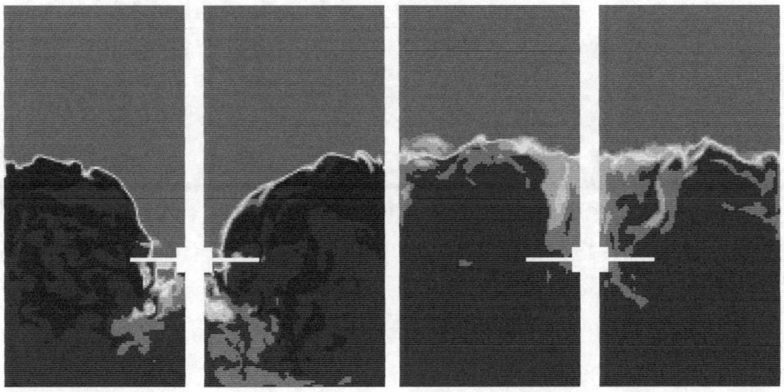

See more of this Session: The Use of CFD In Simulation of Mixing Processes
See more of this Group/Topical: North American Mixing Forum

(657a) Resonant Acoustic (Vibratory) Powder Mixing Performance

Juan G. Osorio and Fernando J. Muzzio, Chemical and Biochemical Engineering, Rutgers, The State University of New Jersey, Piscataway, NJ

Powder mixing is one of the most important operations during food, chemical and pharmaceutical processing. Blending performance of various blender designs, including V-blenders, double-cone blenders, tote-blenders and rotary-drum blenders, have been previously studied in depth. The efficiency of the powder mixing process is measured as the speed at which the desired uniformity of a blend is achieved. The blending performance has been found to be dependant on blender type, material properties and even environmental conditions. In these studies, the performance of a new laboratory-scale ResonantAcoustic® Mixer (RAM) is studied. The RAM uses acoustic energy to provide both bulk and micro mixing. Resonant acoustic mixing is known to significantly reduce mixing time making it a good candidate for improving the efficiency of powder mixing processes.

The blend uniformity of lactose and acetaminophen powders blended using the RAM is compared to mixing results obtained using a double-cone tumble blender. Uniformity is determined by the variance in acetaminophen concentration (as measured by near infrared spectroscopy (NIR)) obtained from numerous small samples collected over the duration of mixing. The optimal RAM operating parameters resulted in relative standard deviation (RSD) values for acetaminophen concentration of 0.4 as opposed to 0.8 for the double cone blender. The RAM achieved these minimal RSD results in 30 seconds as opposed to 240 seconds for the double cone blender. In the second part, a fractional factorial design of experiments is performed to characterize and optimize the mixing performance of the RAM. Variations in fill level, mixing time, and mixing intensity were used. Combinations of binary mixtures were used in order to understand the effects of mixing parameters and material properties. Three fill levels (25%, 50%, and 75%), three mixing times (1 min, 2 min, and 4 min), three mixing intensities (25 Gs, 50 Gs, and 75 Gs) were used as mixing parameters. The binary mixtures used were composed of three API grades (micronized and granulated APAP, and caffeine), three API concentrations (3%, 10%, and 30%), and two grades of MCC (Avicel PH102 and Avicel PH200). After each experimental combination, numerous samples were extracted from the 273-ml vessel in order to determine the mixing performance determined by the variance in the acetaminophen concentration (as measured by near infrared spectroscopy). The results indicated that the RSD decreased with increasing acceleration. For the materials and parameters used, the time of mixing was not significant, although higher acceleration results in better mixing. Temperature increased with increasing fill level, time and acceleration.

See more of this Session: Measurement, Monitoring and Characterization Methods for Particulate Systems Research In Powder and Granular Mixing
See more of this Group/Topical: Particle Technology Forum

(657b) Experimental Indicators of Materials Processing Progress In Mechanical Alloying and Reactive Milling

Priya R. Santhanam, Paul Dupiano and Edward L. Dreizin, Otto H. York Department of Chemical, Biological, and Pharmaceutical Engineering, New Jersey Institute of Technology, Newark, NJ

Mechanical milling is widely explored to prepare advanced alloyed and composite materials. However, it is often unclear how to transfer production from small scale laboratory mills to larger scale devices for practical manufacture of the desired materials. Recently, a simplified approach was proposed, in which the rates of energy transfer to the material from milling tools were calculated for three different mills using discrete element modeling (DEM); these rates were correlated with the averaged experimental rates of increase in the yield strength of an oxide reinforced composite as a function of its milling time. The results were encouraging, but two significant limitations remained. First, experimental validations relied on tangible changes in the yield strength of the milled material. Such changes, common for oxide reinforced composites, could be less pronounced or insignificant for other materials. Second, a steady-state refinement was assumed. The latter assumption needs to be revisited if restitution and friction coefficients change during milling, resulting in altered rates of energy transfer from milling tools to the powder. Therefore, one of the goals of this effort is to enable a validation of the DEM calculations for materials, for which measurements of yield strength are impractical. The second goal is to develop real-time indicators of milling progress achieved in an attritor mill, which do not require recovery of the powder samples. The utility of such real-time indicators is two-fold. First, no labor-intensive recovery and examination of the partially milled powders is required. Second, the instantaneous changes in the powder properties, affecting friction and restitution coefficients can be accounted for in modified DEM calculations. The attritor mill is specifically selected for this study as the device most suitable for large scale manufacturing of mechanically milled materials. Correlations between the rate of power consumption, torque, and rpm of the motor and the powder refinement are examined. Yield strength measurements for the recovered samples are correlated with changes in the measured x-ray diffraction patterns.

See more of this Session: Measurement, Monitoring and Characterization Methods for Particulate Systems Research In Powder and Granular Mixing
See more of this Group/Topical: Particle Technology Forum

Investigation of Reaction Mechanism in Ti/C System: Influence of Mechanical Activation

Ya-Cheng Lin[a], Khachatur V. Manukyan[a], Alexander S. Mukasyan[a]

[a]*Department of Chemical and Biomolecular Engineering, University of Notre Dame, Notre Dame, Indiana, 46556 USA*

Abstract

The influence of mechanical activation (MA) on the stoichiometric mixture of titanium and graphite powders at room temperature under argon atmosphere is discussed. Nano-scale TiC (~500 nm) is produced within 9.5 min of high energy ball milling (HEBM). Crystal structures and phases were characterized by X-ray diffractometry (XRD) and energy dispersive X-ray spectroscopy (EDS). Microstructure of the initial and as-milled samples was done by SEM and internal structures characterized by focused ion beam (FIB) technique. Ignition characteristics and activation energy calculations were carried out by the so-called electrothermal explosion (ETE) method [1]. Special attention was focused on the microstructure variations with respect to ignition temperatures at different milling durations. Reaction mechanism regarding chemical route is also proposed.

First, BET specific area analysis showed ~100 (m^2/g) after 2 min of MA (3 m^2/g for initial mixture). At this stage, microstructure images and XRD patterns indicate the complete amorphorization of graphite powders on Ti bulks. Second, preliminary SEM results reveal the fact that Ti/C laminated microstructures are observed as early as 3 min of MA, corresponding to an evident decrease of ignition temperatures from 1950 K to 1800 K. Third, the appearance of nano TiC_x particles (20–50 nm) at the internal as-milled powder microstructure images depicted by FIB also suggests that evolution of fine grains of TiC_x in terms of milling durations (3–7.5 min). Finally, ETA results showed the ignition temperatures further decrease to ~1500 K and stabilize for samples after 7.5 min MA. The results obtained from microstructure evolution with respect to different milling durations (0–9.5 min) are in good agreement with the decrease of ignition characteristics and other metallurgical analyses.

[1] A. S. Shteinberg, Y.-C. Lin, S. F. Son, A. S. Mukasyan. "Kinetics of High Temperature Reaction in Ni-Al System: Influence of Mechanical Activation" *J. Phys. Chem. A*, **2010**, *114* (20), pp 6111–6116.

Outline for Metal Powder Processing Safe Working Practices

Abdollah Kashani
Safety Consulting Engineers, Inc.
Schaumburg, Illinois 60173

Biography

Abdollah Kashani *a Senior Safety Project Engineer has over 30 years experience providing innovative solutions and services for a variety of industrial, pharmaceutical, medical device manufacturing, and governmental customers. He has been providing services for various processes and operations such as material handling, heat generation, material transfer and transportation, hybrid mixtures (liquid-gas-solid) and other operations.*

Involved in design and development of simulated process conditions for material responses. Examples are electrostatic charge generation, electrostatic discharge vapor/gas/solid explosions under elevated temperatures and pressures, identifying factors that influence the process and related hazard and safety issues.

Provided customized services for qualification of materials and equipment for the end user such as ignition protection study of electrical devices, in flammable environments, simulated to understand the cause of fire or explosion related to particular incident investigations.

On an industrial scale, he and his team developed and designed customized solutions such as flame/detonation arrestors, cyclone for high energetic materials, prototype explosive detection system, and conducting process safety/hazard analysis.

His experience also includes material flammability and fire resistance, demilitarization service, open gas/vapor and dust cloud explosion, hydrocarbon spray and pool fire, electrostatic charge/discharge study, propellant and energetic material burning rate combustion, design and setup for testing of hazardous material for DOT Class 1, 2, 3, 4, 5, 8 & 9 per applicable standard methods, and non-destructive evaluations.

He managed the customized service to be efficient and cost effective with quick turn-around-time. Other expertise included limiting oxygen concentration, flammability envelope, and ignition energy, characteristics, assessing the risk runaway reaction of material at variety of conditions and validation for exothermic and kinetic reaction.

He obtained his Associate Degree in Civil Engineering and Bachelor of Science in Medical Technology and Sciences from Iran. He obtained his Master of Science in Bioengineering from the University of Illinois, Chicago, Illinois. He published jointly a number of technical papers.

Abstract

Work related injuries and accidents have been a continuing problem in various industries. Proper training of workers and understanding the various aspects of a process, hazard conditions through customized audio-visual communication tools can be effective and a perfect way of introducing the employer and employees to safety issues and safe working practices.

The human brain remembers visualized information easier and helps workers to ensure a safe work environment. No one and no work place can afford to have accidents.

In this paper a detailed and thorough outline for safe metal processing operations is being presented. The outline presented here can be used to develop a training course or a safety video which will motivate workers toward safe behaviors, and a greater compliance with safe system working practices.

Identification of hazard conditions, implementation of awareness of hazards in the work place, various prevention and protection measures to ensure safe working practices, methodology to determine hazards, powder behavior and its operation as well as adopting suitable means of controlling hazards are discussed herewith.

Customized training and understanding the safety and hazards of a particular operation and process (e.g. metal processing industries) can eliminate or minimize work related injuries and accidents.

Key words: safe work place practice, metal powder process operation, safety training outline, process safety

Introduction

In the metal powder, coal mining, grain handling, plastics, chemicals and wood products industries, dust explosions, fire, and toxicity have been a continuing problem. The most common sources of safety and health problems in manufacturing are the machinery and its operation. Various preventive and protective measures must be taken to ensure safety.

Purpose

This presentation should be used as a communication tool for the promotion of safe working practices. It alerts employers and employees to the hazards of work environment by pointing out the means of identifying, eliminating and controlling hazard conditions. In this paper the focus will be on the **Metal Powder Processing Operation.**

Scope

How dust explosions can occur in the Metal Powder (PM) Producers and Parts Makers operations, how metal working machinery/tools can fail or operate in an unsafe manner and how workplace injuries can occur? What are the steps and processes necessary to cause dust explosions or to generate toxic environments? Safe practice operations must address preventive and protective measures dealing with:

- Generation of a dust cloud in the PM production and PM parts makers
- Generation of a toxic environment in the PM production and PM parts makers
- Generation of conditions necessary for metal dust explosions
- Failure of machinery, or their operations after unsafe conditions are developed

General Hazards Identification

A) **Fire**

- Extreme fine powder particles when exposed to air could react with air and catch fire.
- A powder layer can be ignited by the following ignition sources:
  - Electrostatic discharge (ESD)
  - Heat
  - Impact
  - Friction
- Metal powder generates high heat flux and can lead to metal powder explosions (and secondary explosions).

B) **Explosion**

- Fine particles (with high surface area/volume) can be lifted up and create an explosive dust cloud.
- A dust layer can be kicked and lifted up to create an explosive dust cloud.
- Metal powder explosions can be initiated by various ignition sources such as:
  - Pyrophoricity
  - Electrostatic discharge (ESD)
  - Heat
  - Impact
  - Friction
- Dust explosibility characterized by pressure output/rate of pressure rise.
- Hybrid mixture of metal powder and flammable liquid will cause higher pressure output/rate of pressure rise.

C) Workplace Injuries

Workplace injuries, which can occur due to lack of safeguarding and safe operation, are mostly related to:

- Improper ventilation
- Improper design or construction
- Improperly or inadequately guarded or unguarded machinery
- Exposed moving parts
- Sharp cutting edges
- Power presses and power hammers
- Furnace and exhaust systems
- Defective tools and/or equipment
- Vibration and noise from machinery
- Physical reaction either voluntary or involuntary

Types of injuries could be:

- Permanent total disability (including death)
- Permanent partial disability
- Temporary total disability

D) Health and Toxicity

Health hazards that pertain to fine metal powders/vapors which may enter the body via the following:

- Inhalation
- Skin contact (dermal)
- Ingestion (oral)

Health and Toxicity due to exposure can cause:

- Lung damage (breathing)
- Burns/corrosive burns
- Skin reaction due to contact with metal
- Dust (allergic reaction)
- Interference with normal metabolic function

Process Operation and Hazards

Process Operations:

The main process operations in metal powder and/or parts production are:

A) **Powder/Part Production**

- Atomization
- Chemical reduction
- Spray drying of metal powder
- Milling/grinding
- Pressing operations
- Sintering
- Injection molding
- Parts (machining)secondary operations

B) **Powder/Part Handling**

Major preparation operations for either metal powder and/or parts are:

- Screening
- Cleaning
- Annealing
- Mixing and blending
- Heat treating
- Lubricating

C) **Metal Powder Press**

Mechanically or hydraulically powered machines are:

- Metal sizing press
- Metal coining press
- Compacting press
- P/M briquetting press
- Steel and plate press
- Turning machines
- Boring machines

Hazards:

The hazards associated with any of the above operations are as follows:

A) **Dust Explosion**

- Presence of ignition sources due to:
  - Heat
  - Atomization
  - Grinding and milling
  - Annealing, sintering

- Electrostatic discharge due to:
  - Movement of powder
  - Mixing and blending
  - Screening

- Electric sparks due to:
  - Failure of electrical circuitry
  - Welding operations

- Friction and impact due to:
  - Grinding
  - Milling
  - Parts cleanup
  - Screening

- Thermal effect due to:
  - Annealing
  - Sintering
  - Drying
  - Heat treating

B) **Gas/Vapor Explosion**

- Generation of hydrogen (water reactive metal powder e.g. aluminum)
- Release or leak of hydraulic fluid
- Cleaning Solvent Vapor

C) **Mechanical Hazard**

- Ejection of hard object from presses, grinding, milling, turning, etc.
- Moveable and/or rotating machinery
- Failure of electrical circuitry
- Failure of interlock or interlock bypass
- Inadequate guard, guard removed or barrier guard not provided
- Slippery or rough surfaces
- Impact by moving equipment or parts
- Toppling or overturning of unstable products
- Excessive weight
- Excessive vibration and noise

D) **Toxicity Hazard**

- Generation of toxic dusts
- Generation of release of toxic metal fumes

- Release of toxic vapors/fumes from chemical reactions or processes
- Inadequate oxygen level for breathing
- Direct contact with metal dust

Guidelines to Mitigating Hazards

Hazards can be avoided by proper design, control, procedures and training.

A) **Explosion**

- Minimize powder handling
- Keep the flow of the powder at a low rate
- Use dust collecting/filtration system
- Minimize generation of fine metal powder particles
- Prevent dust accumulation
- Prevent air leak into furnace/equipment
- Removing ignition sources such as ESD, heat, friction, impact, etc.
- Use explosion venting or explosion suppression system
- Housekeeping (maintenance)

B) **Injury Prevention**

- Machines should never be left running unattended unless the machine has been designed to do so.
- Operators should not wear jewelry, loose fitting clothing or long hair, which could be caught in moving parts.
- Use personal protective equipment (PPE), such as respirators, gloves
- Manual adjustment, gauging of work should not be practiced while machine is in operation.
- Metal chips or dusts should not be removed by hand. Use brushes, vacuum equipment or special equipment for removing chips or dust.
- Use only the proper hand tools for each job or machine
- Do not use high-pressure compressed air to blow off dust or dirt. Use a vacuum system instead, whenever possible.
- Machine guards, baffles, or chip guards should be around the machinery to shield the operator.
- Select equipment for installation which has low vibration and noise.
- Equipment which could vibrate should be mounted on firm, solid foundations and secured with bolts.
- Provide sound absorbing enclosures or vibration insulators for noisy or vibrating equipment.
- Safety guards such as, anti-repeat, barrier gate devices, stop control, brake control and spring guards should be used.

Methodology for Example of Dust Explosion Hazard Testing and Analysis

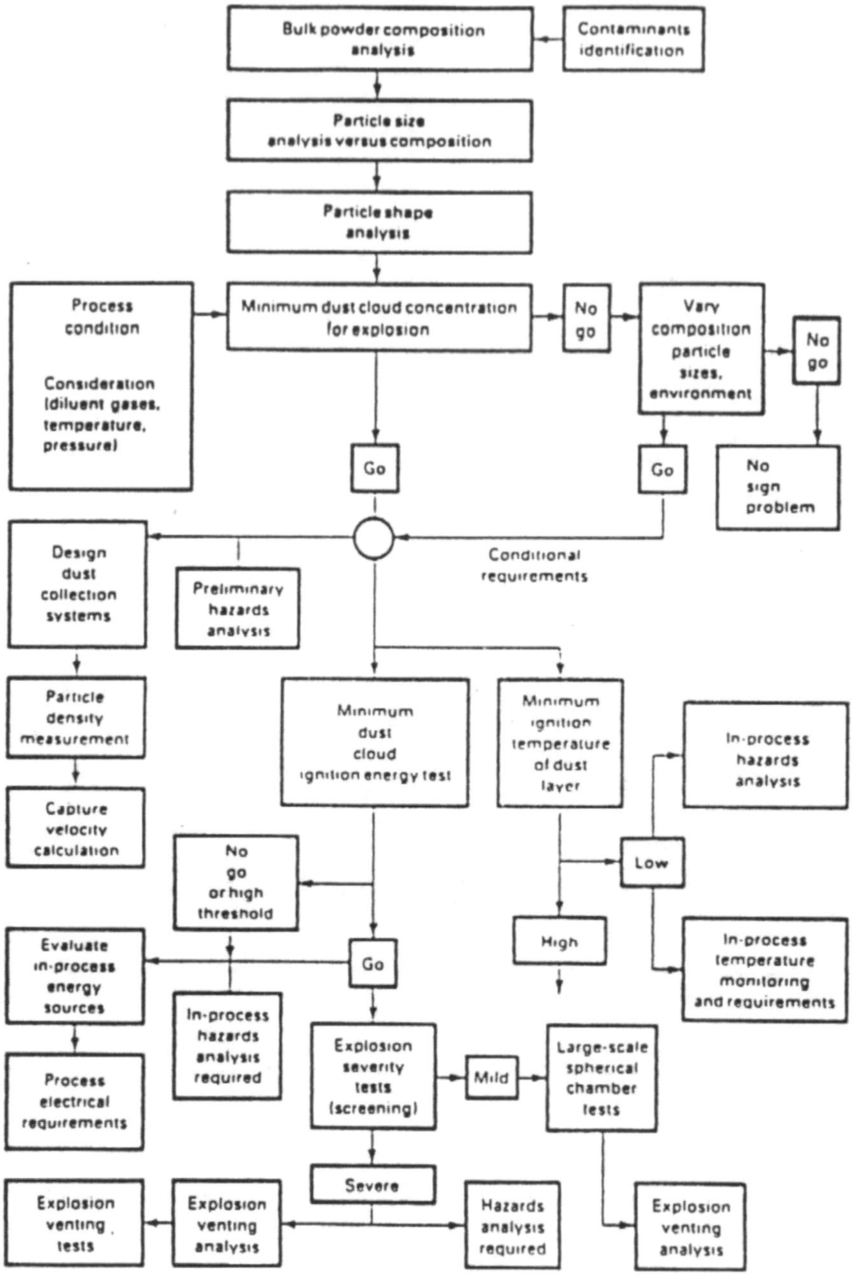

Material dust explosion hazard test methodology.

Table 1

Example of Ignitability and Explosibility of Metal Dust

Dust Type	Minimum Explosible Concentration (gm/m^3)	Output Pressure (barg)	K_{St} (bar·m/sec)	Minimum Ignition Energy (mJ)
Fine Aluminium Type 1	30	11.5	1100	2-3
Fine Aluminium Type 2	60	10.2	100	10-15
Iron Powder, Fresh	500	5.2	50	--
Aluminium/Iron (50:50)	250	9.4	230	--
Fine Silicon Powder	125	10.2	126	3
Zinc Powder DC1	125	7.3	176	2
Zinc Powder DC1	250	6.8	93	--

Conclusion

Each product and design operation may have certain inherent hazards. Proper training through communications will allow employers/employees to eliminate or control hazardous conditions by:

- Identifying hazards that could cause fire, explosion and toxicity so corrective actions and any necessary changes can be implemented, minimizing personnel exposure
- Adopting protective measures means incorporating suitable safety measures in the process design for specific hazards encountered during the operations.
- Contributing to a safe working environment and safe operations means providing accurate safety procedures and standards
- Reporting any hazardous conditions or defective equipment that may exist.
- Provide personnel training on a regular basis.

A financial crisis is more likely to occur when there is an accident, because firms find that their real burden of indebtedness is increasing while there is no increase in the production and in the real value of their assets. The resulting decline in the firm's net worth increases. No one and no work place can afford to have accidents.

A great opportunity to explore a wide variety of smart, safe, sustainable manufacturing solutions tools and programs are available such as:

D) **HAZID** Hazard Identification Study
E) **PHA** Process Hazard Analysis
F) **FMEA** Failure Mode and Effect Analysis
G) **HAZOPs** Hazard and Operability Studies
H) **HFMEA** Health Failure Mode and Effect Analysis
I) **HACCP** Hazard Analysis and Critical Control Points
J) **PHA** Proactive Hazard Analysis
K) **RCA** Root-Cause Analysis
L) **LOPA** Layers of Protection Analysis
M) **TQM** Total Quality Management

These are some solutions which are geared toward improving your safe working practices and your business profitability.

References

Baker, Wilfred E., "Explosions in Air" 1973, University of Texas Press, Published in cooperation with the Southwest Research Institute

Bartknecht, Wolfgang Dr., "Explosions, course prevention protection" 1981, Springer-Verlag Berlin Heidelberg New York 1980

Dahn, C.J., "Explosivity and Pyrophicity of Metal Powders" Metals Handbook, ninth edition. Volume 7 – Powder Metallurgy.

Eckhoff, Rolf K., "Dust Explosions in the Process Industries" third edition 2003, Gulf Professional Publishing

Sax, N. Irving, "Cancer Causing Chemicals" assisted by Elizabeth K. Weisburger, David Schottenfeld, Joanna Haas, Benjamin Feiner, Barry I. Castleman and Richard J. Lewis, Sr., Copyright 1981 by Litton Educational Publishing, Inc. Van Nostrand Reinhold Company

(657e) Experimental and Numerical Study On the Tablet Velocity In Commercial Scale Pan Coaters

Wei Chen, Shih-Ying Chang and San Kiang, Dpst, Bristol Myers Squibb, New Brunswick, NJ

Effective control of coating uniformity in a tablet coating process requires good mixing. Scaling up a pan coating process from lab to commercial scale to maintain the same level of tablet uniformity poses unique challenge. It was found in our previous study that [1] the tablet flow rate, a value calculated from the tablet velocity number density (tablet size) and spray zone width, can be an indicator of the mixing condition. We have also quantified the impact of the tablet velocity and number density on the tablet content uniformity. In this paper, the tablet velocity in commercial pan coaters was investigated experimentally and numerically. This study was carried out in 50kg, 400kg and 600kg pan coaters. Different types of tablets including oblong and round tablets with weights from 200mg to 1500mg were used in the study. A real-time imaging device was developed to measure and analyze the tablet flow behavior within the sprayed area. The current study shows that the smaller round tablets move faster than larger ones in general. However, for the oblong shape tablet the larger tablets move faster than the smaller ones. For all the tested tablet types, the tablet velocities increase with increasing air flow rates. This enhancement by air flow starts to level off at the higher air flow rates. For a given pan size, there appears a pan loading that has the maximum tablet velocities. The existence of such maxima implies an optimal pan loading at which the desired coating uniformity can be consistently obtained. In the parallel study through the numerical simulation, a Discrete Element Method (DEM) was applied to simulate the tablet flow behaviors in a pan coater. The simulation results show a good agreement with the experimentally measured tablet velocities.

See more of this Session: Measurement, Monitoring and Characterization Methods for Particulate Systems Research In Powder and Granular Mixing
See more of this Group/Topical: Particle Technology Forum

(657f) Determination and Modeling of the Limiting Oxygen Concentration of Pesticide Dusts with Highly Corrosive Combustion Products

Hannes Kern and Harald Raupenstrauch, Chair of Thermal Processes, University of Mining Leoben, Leoben, Austria

Within the production of agricultural pesticides various different kinds of dusts occur which can be considered as explosive due to their chemical and physical characteristics. In various steps of the production process hybrid mixtures (e. g. dust in acetone) are present. For most of the considered dusts with high contents of halogens there are no sufficient data available in order to determine the explosion hazardous potentials of the different process steps. In this work safety characteristics, especially the limiting oxygen concentration, of 2,4- Dichlorophenoxyacetic acid- dimethylamine (2,4D- DMA) and different urea herbicides (Chlortoluron, Fluometuron) have been determined experimentally and theoretically.

To characterise the occurring dusts also not only in terms of their chemical composition extensive research on the particle structure and surface as well as on the particle size distribution were carried out. The particles show a crystal structure with no porosities or inclusions. The parameters of the particle size distribution were approximated by a RRSB distribution. All dusts show values around 1 for the scattering parameter n. Besides the particle size distribution also the burning behaviour of the dusts was investigated. For all dusts no autonomous burning could be observed.

After research of the dusts under a microscope and the analysis of the particle size distribution and burning behaviour, experiments in the ISO 20l apparatus of the company FireX were carried out. All observed dusts show significant values of pressure rise and maximum explosion pressure. The maximum explosion overpressures are situated around 8bar and the measured minimum ignition energies lie between 8 and 16mJ.

The measured values for all analyzed dusts where also compared to results of different models for calculation of safety relevant characteristics.

A main goal of the present work was to determine the limiting oxygen concentration of the chosen dusts. In the course of the research the combustion products of the dusts turned out to be highly corrosive, which can be explained by the chemical composition of the dusts and the burning conditions in the ISO 20l apparatus. Some combustion products caused smaller damage on the 20l apparatus, due to corrosion effects. A reliable method to measure the limiting oxygen concentration had to be found as alternative for the 20l apparatus.

By modification of the Hartmann apparatus, which is used for screening purposes, a method was to determine the limiting oxygen concentration (LOC) was investigated. A goal of the chosen method was also to reduce the amount of sample needed for the analysis. The method was first checked by several dusts with well known values of limiting oxygen concentration. All values for the LOC of the investigated dusts, could also be found by this alternative method in consideration of the defined failure limit. The needed sample amount for the analyses is significantly lower compared to the ISO 201 apparatus.

All results of the test dusts were also compared to results of several different models for the determination of the limiting oxygen concentration. The models showed good agreement with the values found in experiment, so that the method could be considered as reliable under the chosen experimental conditions.

As a next step all pesticide dusts where investigated with this alternative method. Compared to all other safety relevant characteristics determined in the course of this work, also the limiting oxygen concentrations of the dusts showed comparable behaviour, regarding their chemical composition. All results of the experiments where compared to results of mathematical models and showed good agreements. The determined limiting oxygen concentrations where also comparable to other known dusts with quite similar chemical composition.

The used method offers the possibility to determine the limiting oxygen concentration of dusts with highly corrosive or hazardous combustions products by limiting the amount of sample needed and simplifying of the cleaning process of the equipment due to the assembly of the used Hartmann apparatus.

See more of this Session: Measurement, Monitoring and Characterization Methods for Particulate Systems Research In Powder and Granular Mixing
See more of this Group/Topical: Particle Technology Forum

AUTHOR INDEX

AUTHOR INDEX

American Institute of Chemical Engineers
3 Park Avenue
New York, New York 10016-5991

CURRAN ASSOCIATES INC
proceedings.com

2011 Annual Meeting of the American Electrophoresis Society

(AES)

Topical Conference at the 2011 AIChE Annual Meeting

Minneapolis, Minnesota, USA
16-21 October 2011

AIChE

ISBN: 978-1-61839-577-1